The Human Gene Editing Debate

26 déc 2020-29 déc 2020

Conley and Fletcher, The Genome Factor
Buchanan et al., From Chance to Choice
(Rawlsian)

The Human Gene Editing Debate

JOHN H. EVANS

OXFORD
UNIVERSITY PRESS

OXFORD
UNIVERSITY PRESS

Oxford University Press is a department of the University of Oxford. It furthers the University's objective of excellence in research, scholarship, and education by publishing worldwide. Oxford is a registered trade mark of Oxford University Press in the UK and certain other countries.

Published in the United States of America by Oxford University Press
198 Madison Avenue, New York, NY 10016, United States of America.

© Oxford University Press 2020

All rights reserved. No part of this publication may be reproduced, stored in a retrieval system, or transmitted, in any form or by any means, without the prior permission in writing of Oxford University Press, or as expressly permitted by law, by license, or under terms agreed with the appropriate reproduction rights organization. Inquiries concerning reproduction outside the scope of the above should be sent to the Rights Department, Oxford University Press, at the address above.

You must not circulate this work in any other form
and you must impose this same condition on any acquirer.

Library of Congress Cataloging-in-Publication Data
Names: Evans, John Hyde, 1965– author.
Title: The human gene editing debate/John H. Evans.
Description: New York, NY, United States of America : Oxford University Press, 2020. |
Includes bibliographical references and index.
Identifiers: LCCN 2020002559 (print) | LCCN 2020002560 (ebook) |
ISBN 9780197519561 (hb) | ISBN 9780197519578 (updf) |
ISBN 9780197519585 (epub) | ISBN 9780197519592 (online)
Subjects: LCSH: Gene editing—Moral and ethical aspects. | Human genetics.
Classification: LCC QH438.7 .E933 2020 (print) |
LCC QH438.7 (ebook) | DDC 599.93/5—dc23
LC record available at https://lccn.loc.gov/2020002559
LC ebook record available at https://lccn.loc.gov/2020002560

1 3 5 7 9 8 6 4 2

Printed by Integrated Books International, United States of America

Contents

Acknowledgments	vii
1. Introduction	1
The Importance of Public Bioethical Debate	5
Slippery Slopes	9
The Slope as the Micro-Structure of the HGE Debate	11
Slippery Slopes in the HGE Debate	15
Barriers	18
2. The First Barriers in the Human Genetic Engineering Debate	25
The First Barriers	29
Why These Barriers Were Strong	35
A Change in the Participants and Their Values	44
Weakening the Somatic/Germline Barrier	46
Weakening the Disease/Enhancement Barrier	57
The Somatic/Germline and Disease/Enhancement Barriers at the End of the 20th Century	65
3. The CRISPR Era, the National Academies Report, and the Median Trait Barrier	67
The Barriers at the Beginning of the CRISPR Era	69
The NASEM Report Advocates Replacing the Somatic/Germline Barrier	77
The NASEM's Proposed Barrier	85
The Prevalent Variant Barrier	87
Design Strength of the Barrier	90
Future Polygenic Selection and Modification	94
Designing the Median Trait Barrier	97
Values That Support the Median Trait Barrier	100
Values That Imperil the Median Trait Barrier	103
The Barriers Examined in This Chapter	108

4. Possible Barriers Further Down the Slope	111
The Goals of Medicine Barrier	113
The Family Genes Barrier	115
The Boundary of Humanity Barrier	122
The Liberal Eugenics Barrier	126
Conclusion	131
5. Conclusion	133
Historical and Future Barriers on the HGE Slope	134
Can Any Barriers Hold?	140
Could Solid Barriers Be Anchored by Authority?	142
Rejecting the Direction of the Slope	143
General Conclusions About Slippery Slopes from the HGE Case	145
What Is to Be Done?	150
Notes	153
Bibliography	181
Index	191

Acknowledgments

My first academic interest was in what is now called the human gene editing debate. Nearly 25 years ago, I started a dissertation on the topic, which eventually resulted in my first book. Not too long after, I concluded that the most interesting component of that debate—human germline modification—would remain scientifically impossible. I moved on to related questions. That was premature, because within 10 years scientists had developed the CRISPR gene editing technology that appears to make the impossible possible.

My interest in the debate was re-kindled by joining the Committee on Human Gene Editing: Scientific, Medical and Ethical Considerations convened by the National Academies of Sciences, Engineering, and Medicine. I greatly enjoyed my fellow committee members, and their views helped me sharpen my thinking about the nature of this debate. Needless to say, this book represents my own views and not those of the other committee members or the National Academies.

I appreciate comments and discussions about earlier drafts with Lisa Madlensky, Ellen Clayton, Matthew Porteus, Dan Navon, Dalton Conley, and reviewers from Oxford University Press. I also benefited from audiences at the Center for Applied Philosophy and Ethics at Kyoto University and the UCLA Institute for Society and Genetics. The UC San Diego Sociology Department and Institute for Practical Ethics provided a collegial incubator for ideas. As usual, the Science, Technology and Innovation Studies unit at Edinburgh served as a supportive home away from home. And, as always, my spouse Ronnee was always willing to help with my ideas and served as an endless source of support and love.

The Human Gene Editing Debate

1

Introduction

On November 26, 2018, the news broke that a Chinese scientist had facilitated the creation of twin girls who supposedly had been genetically modified to make them less susceptible to the HIV virus.[1] If the claim ends up being true, it would be the first attempt to directly change the genes in a human in a way that would change all of their progeny. It was also the first attempt to make such a change not to fight a genetic disease but for what would broadly be considered a human enhancement—an improvement in not only those girls but ultimately also the human species. A firestorm of controversy erupted. By the end of this book I hope you will see why. This act was a long leap over a number of established barriers on the slippery slope—the theoretical metaphor that grounds this book—which landed genetic modification in a zone where scientists in an earlier era said they would not go.

This book is broadly about the debate concerning human genetic engineering (HGE), which I define as intentionally changing the genes in a human from what they otherwise would be. There have been many methods used and proposed for HGE over the decades, which I will cover in the upcoming pages, but the most recent method is the subject of a lot of conversation: human gene editing. I am primarily focused on the US version of this debate.

To give more context about the ethical debate about HGE, we must briefly go back to discussions about human genetic inheritance in the 19th century. I am sure it was obvious to people for thousands of years that tall people who mate with each other tend to produce tall children, and people with dark skin tend to produce children with dark skin. By the 19th century an Austrian monk named Gregor Mendel, experimenting with cross-breeding pea plants, figured out the fundamental laws of inheritance, such as the fact that (what we now call) genes come in pairs and one is inherited from each parent. But how exactly does that work, *and*, critically, can it somehow be controlled? All sorts of social engineers have long viewed genetics as a tool to improve society.

The first of these social engineers were the eugenicists. "Eugenics" means the science of improving a human population by controlled breeding,[2] and the leaders of the eugenics movement of the late 19th century saw that the rich tended to have successful children and the poor less successful children. In fact, not only physical health but also intelligence, moral behavior, and personality traits were thought to be the result of proper breeding. Whereas today we would probably conclude that a lack of success is the result of the social and cultural transmission of inequality, back then the conclusion was that the rich had better genes. If society could stop those with "inferior" genes from reproducing and encourage the reproduction of those with "superior" genes, not only would social problems be lessened but the human species could also be improved.

As much as it is easy to ridicule the scientific and social ignorance of the early 20th-century eugenicists, they ended up being right about a few of the basics. There are genes that are passed down through reproduction that result in diseases that cause great suffering, like sickle cell anemia. There are also genes that could, if somehow inserted into an individual's genome, make them "better" than they otherwise would be, and if these modified persons reproduced, they would slowly "improve" the human species. (The genome is the complete set of genes in a cell or organism.)

The eugenicists advocated selective breeding for humans, as has long been done with farm animals. To influence human reproduction, various policies were put in place in the United States and Europe, such as forced sterilization of those perceived to be genetically inferior, restricting immigration from countries whose people had "undesirable" genetic qualities (and thus preventing potential mating and offspring), and generally encouraging "better" humans to have more babies through moral suasion and propaganda.[3] There was very little understanding of what was actually a heritable trait and what was just a reflection of the social biases of genetic scientists. After all, this was an era when it was thought that some people's genes made them "moral imbeciles."

In this era and into the 1940s, what came to be called the "mainline" eugenics movement assumed that positive and negative traits were structured by race and class. This logic reached its height with the Nazis, who presumed that the "Aryan race" had superior genetic traits and others had negative genetic traits. This thinking provided the scientific justification for the Holocaust, when millions of people from "genetically inferior" races were killed.

This book is about the debate over genetically changing humans that started in the 1950s, after the discrediting of coercive race-based eugenics that reached its apotheosis with the Nazis. Scientists still had little idea of how genetic inheritance actually worked, but this changed with the 1953 discovery by Francis Crick and James Watson of the chemical structure of deoxyribonucleic acid—DNA. It turned out that the elementary units of inheritance—by then called genes—were just particular sequences of chemical bonds. This reinvigorated the chastened eugenicists, who still nurtured the dream of creating an improved species, albeit now through voluntary, racially neutral eugenics.

Yet, no longer would advocates have to develop various mating schemes to get the "right people" to have more babies—if the genome is merely chemicals, "improved" genes could just be inserted. Scientist Robert Sinsheimer wrote during this era that the new technologies allowed for "a new eugenics. . . . The old eugenics would have required a continual selection for breeding of the fit, and a culling of the unfit. The new eugenics would permit in principle the conversion of all of the unfit to the highest genetic level . . . for we should have the potential to create new genes and new qualities yet undreamed" in the human species.[4] As one would expect, there was an ethical debate at the time about whether we as a society should engage in HGE.

I will fill in much detail in the following pages, but for now I will jump forward to today. Knowledge of what particular genes do for the human body has increased dramatically, but the ability to correctly insert a gene into an individual's genome has remained elusive. For many years, scientists tried to heal people who had genetic diseases by getting a proper version of the gene in question delivered to the malfunctioning cells in their body. The delivery vehicle was a virus, which itself sometimes caused problems, and, more importantly, often delivered the gene to the wrong place on the genome, causing another set of problems. To overgeneralize, from the 1960s to the 2010s, HGE would not work. Unable to even heal genetic diseases in people, Sinsheimer's dream of improving the human species remained fanciful.

The debate about the ethics of HGE lost energy as it appeared that the most controversial applications of this technology would be impossible. I contributed, in a very small way, to the decline in the debate. My first book, published in 2002, was a sociological explanation of the ethical debate about HGE up to 1995.[5] Shortly after publishing that book, I concluded that the more controversial applications of HGE would always be impossible, and I moved on in my career to related topics I thought more relevant.

My decision was premature. In the early 2010s, scientists invented a new gene delivery system called CRISPR-CAS9. It removes the unwanted part of a gene and then places a new part where it is supposed to go on the genome. How CRISPR works is not important to this book. All you have to know is that it is already being used in humans.[6] This precision is reflected in the desired metaphor of the scientists: human gene "editing." Whereas the old techniques were like randomly inserting a word on a manuscript page, while hopefully not disrupting the flow of the text, the new technique is like a word processor search and replace function.

Editing the genes of people with single-gene recessive diseases like sickle cell anemia seems immanent, and the eugenicists' dream of creating improvements in the species not impossible. The development of the CRISPR system has started the scientific race to be the first to efficiently "edit" human gametes, which has supposedly already happened in China.[7]

After decades as a bit of a backwater, the debate is back on, and it has the feel of the HGE debate of the 1960s, complete with Nazis. It is not usually a scientist who starts talking about Nazis in these debates, but the gravity of the situation is reflected in the fact that a scientist central to the development of the CRISPR system, Jennifer Doudna, revealed to a reporter she was having nightmares where Hitler would ask about her technology so he could use it for his goals.[8] For the first time, scientists have the means and the desire to conduct the most controversial types of HGE, thus the debates have returned with a vengeance.

The Importance of Public Bioethical Debate

This book is not about any one individual's ethical decisions, but it is about the public ethics of HGE. That is, should we as a society allow HGE? To start with the least controversial endpoint of the debate: Should we allow scientists to use gene editing to replace disease-causing variants of genes so that they are not passed on to the person's offspring? An example would be sickle cell anemia. At present, the social consensus is yes, and in Western countries this is allowed largely under regimes of medical experimentation.

Let us move to the other endpoint. How about allowing people to genetically modify their children, and all of their subsequent descendants, so they have better memories and attention spans? At present that is not possible and would not be allowed in any country in the West that actually has a policy about HGE.[9] But, this position has many advocates in the public debate. This public debate exists among policymakers, among the general public, and, in what I will focus upon, academic ethicists. This is called public bioethical debate, with bioethical meaning concerning the human body.

From my last example you can see that many of the technological possibilities I discuss are speculative. Some scientists will say impossible. However, I think it is useful to discuss the ethics of the seemingly impossible for two reasons. First, it is usually not wise to bet against human ingenuity and the development of knowledge. I had left the debate about germline modification thinking it would always be impossible, and then within 10 years it was possible. Similarly, identifying polygenic traits in embryos was impossible, and then it was possible. Second, even if editing polygenic traits like "intelligence" lies far in the future, or will always be beyond the horizon, it is important for societies to talk about what their collective moral values are. That is, saying "we as a society will never go beyond point X," even if point X remains technologically impossible, is important for a society's self-definition. It also allows scientists to be confident and plan their research knowing what the limits are. Like others in this debate, I think that the benefits of speculating outweigh the risks.[10]

Others will object to this book on the grounds that there is no stopping HGE, so it is pointless to discuss how various limits on HGE are justified. Bioethicist Arthur Caplan writes that "genetic engineering of human beings is inevitable," and "once gene editing becomes a reliable tool to help prevent disease and other undesirable traits, parents might actually be considered negligent if they chose not to use these tools." Moreover, he says, "this idea that we're not going to do gene editing when it gets close to enhancement or improvement, I find it silly and head-in-the-sand kind of stuff."[11] Perhaps, but even this sociologist is struck by the extremity of Caplan's social-structural determinism where resistance is futile. If inevitable, this book can be seen as an explanation for how the inevitable creation of a subpopulation of enhanced humans will come to be ethically justified.

But again, why should we care about public bioethical debate—about what a bunch of university professors think about ethics? The reason is that public bioethical debate has a fairly strong influence on what actually happens. Most scientists do not want to be branded as a "rogue scientist" ignoring the ethical concerns of the society within which they operate. Scientists will often justify their experiments by pointing to the public bioethical literature—particularly seminal reports by government agencies or learned bodies. For example, that Chinese scientist justified his actions by referencing a recent public bioethical report from the National Academy of Sciences, Engineering, and Medicine.[12] (People involved with that report would say that he misinterpreted the report.)

Public bioethical debate also influences public policy, albeit in a diffuse manner. These academics not only produce turgid academic articles, but they also write op-eds for newspapers, are interviewed on television, and testify before legislators. For example, a bioethicist who helped lead recent ethical discussions about gene editing later testified about the technology before the US Senate.[13] The best way to describe the role of this debate is that it has an agenda-setting or framing effect on what actually happens with policy, creating the ethical distinctions that end up circulating in public and among policymakers.

This book does not advocate a particular ethical stance concerning HGE or examine the legitimacy of those ethical positions. I do not argue that a specific application of HGE is wrong or should be illegal. I leave that to philosophers, bioethicists, and theologians, and a quick internet search will get you all the texts you might possibly want to read on the topic.

As a sociologist I have a different perspective. This book is not saying that a certain position is ethical; it instead explains what positions have been and will be *considered to be* ethical in the debate. I take no position on whether those ethical positions that rise to prominence are "good" or even whether they make any sense.

This basic sociology-of-knowledge perspective holds that the quality of the argument is only a component of what results in the argument being accepted—the social context of the argument plays a much larger part. Analogously to Darwinian mutations, there is an endless diversity of ethical arguments available to scholars, but like biological evolution, the environment selects those that best fit with the social context. That is, scholars make all sorts of ethical claims, but only some are accepted; and that acceptance not only has to do with the inherent quality of the argument but also the social status of the claimant and particularly the social power of the ideas they use. To take an extreme example of how the power of ideas changes with social context: No matter what your argument about genetics, if it relies on Protestant theology, it would more likely be accepted in 1918 than 2018, and in the United States rather than in China.[14]

So, what leads an ethical argument to be accepted in a debate beyond its internal wisdom or coherence? For example, and to anticipate what I write later in the chapter, a long-standing ethical distinction is between HGE to change an existing person and HGE to change someone's offspring. Why is that a powerful distinction in this debate? The answer is that, in general, ethical ideas need to be consistent with the beliefs of those with influence in a society.

I have previously written a macro-analysis of how and why the debate about HGE changed from roughly the 1950s to the end of the 20th century, and I cover much of that history in the early part of the present book.[15] However, the present book is not about those macro-forces that shape the debate but about the micro-structure of the debate itself—the limited set of concepts that have long been debated. And, this micro-structure has a very large influence on what can happen in the debate. Let me provide an analogy.

Imagine that you have decided that the US Congress does not represent the views of the citizens of the United States, and you have a plan for remedying that by making voting power in the Senate proportionate to the number of citizens in each state. To account for the different populations in states, a senator from Wyoming would get 1 vote and a senator from California would get 80. You start a social movement to enact these changes. Your movement of citizens is analogous to my macro-force—such a movement is a reaction to macro-forces like income inequality, urbanization, the growing power of corporations, and so on. However, your movement will not be able to do exactly what you want, but you are going to do the best you can *within the confines* of that highly constraining micro-structure called the US Constitution. Yes, it is possible to have a revolution or amend the Constitution—and it is possible to have an analogous revolution in public bioethics—but the odds of that happening are low.

The US Constitution limits your actions to a subset of what you want to do, and the point of this book is to show how the micro-structure of the debate about HGE itself will limit the ethical positions that could plausibly be taken in the future. For example, you could say that HGE should only be allowed for diseases, but, as I will argue in the next chapter, the micro-structure of the HGE debate will make that an untenable position. Your time is best spent on a different argument.

Slippery Slopes

The primary micro-structure of the public bioethical debate about HGE is the slippery slope. The term comes from the slippery slope argument, which starts with a metaphorical slope, with the most meritorious action at the top, and at the bottom is the action that is maximally objectionable from the view at the top.[16] If we step onto the top of the slope by selecting the desirable option of A, then it is *more likely* in the future that we will select the currently undesirable option of B, somewhat further down the slope. Arriving at B, we are then *more likely* to select an even less desirable option of C. At the top of the slope is the initially morally worthy decision, but that decision changes the context of the next decision, making the undesirable choice below more likely. Eventually we are at the bottom of the slope where we did not want to be when we took the first step, where maximally objectionable actions lie.

Probably the most well-known contemporary slippery slope argument is that allowing physician-assisted suicide tends to lead to voluntary euthanasia, which in turn will tend to lead to involuntary euthanasia—the act that most people consider morally objectionable. There is an ongoing debate over whether Belgium and the Netherlands are currently in the midst of this slide down the slope.[17]

What scholars call "logical" or "conceptual" slippery slope arguments have a deservedly bad reputation. These hold that once we have allowed A we *must* or *will* allow B, negating the possibility that in the future, near the decision point for B, people might change their mind or social conditions may change. For example, it is a bad argument to say that if we have approved euthanasia for immanently terminal illnesses we have effectively also approved euthanasia for those deemed to lack social usefulness.

On the other hand, what scholars call "empirical" slippery slope arguments are largely considered legitimate, albeit difficult to make, because they involve making probabilistic predictions of the future.[18] A legitimate empirical slippery slope argument needs to identify the social mechanisms that will, in the future, result in an increasing likelihood of selecting B.[19] In law professor Eugene Volokh's terms, we need to think of "the entire range of possible ways that A can change the conditions ... under which others will consider B."[20] These social mechanisms are, to continue my metaphor, the grease on the slope.

To further understand the slippery slope, we can imagine a number of mechanisms that would cause a *society* to slip down the HGE slope. For a legislative mechanism, imagine that in order to engage in HGE to heal the disease of an existing person (the top of the slope), Congress creates a legal framework that limits the ability of those being experimented upon to sue the experimenters. This legal framework would make it easier to take the next step down the slope, and thus is a slippery slope mechanism.

Similarly, a "technical knowledge" slippery slope mechanism is clearly already operating. Near the top of the slope we learn how to modify any genetic sequence in a cell. This increased knowledge learned from step A makes it more probable we will have the knowledge to take step B. The knowledge from step B makes it more likely to take step C and so on.

The most well-recognized societal-level mechanism is the "attitude-altering slippery slope." In this mechanism, "as one act becomes less and less the exception that has to be argued for as acceptable in a particular case, a social climate of opinion sets in which makes it easier for the next stage to arise as a real possibility. As what is tolerated in a society changes, the possible becomes the plausible."[21]

The attitude-altering slippery slope has clearly already been occurring with HGE. As I will discuss in the next chapter, in the 1990s the first genetically modified people were produced. They were modified in a way so that the change would not be passed on to their descendants, and the goal was to treat a severe genetic disease within those existing persons. Essentially everyone in the debate concluded that the risks of taking this first step on the slope was worth any future risk of slippage. But, this first step normalized the idea that a person could be genetically modified, increasing the likelihood of taking the next step down the slope.

The Slope as the Micro-Structure of the HGE Debate

The previous section was designed to get us thinking about slippery slopes, but now I turn to the topic of this book, which is the micro-slippery slope analysis of the public bioethical debate itself. Public bioethical debates are generally set up as *slopes* with the most acceptable at the top and the abhorrent at the bottom. (I will get to the *slippery* part in a few pages.) For example, the ethical acceptability of abortion is structured as a slope, with the terrain of the slope typically being defined by fetal age and the reasons the woman wants an abortion. At the top of the slope is abortion to save the pregnant woman from immanent death, and at the bottom is infanticide for no stated reason. Most people want to be at the top but do not want to get to the bottom. Similarly, the ethical acceptability of euthanasia has at the top euthanasia for fully competent adults who are going to imminently die a painful death due to disease, and near the bottom is involuntary euthanasia for those people authorities conclude should no longer live.

There are a number of reasons why public bioethical debates about science, medicine, and the human body tend to be structured as slopes. First, the origin of public bioethical debate *itself* was in the overreach of scientists and physicians, and the public calling for limits to avoid a dystopian future. Scandals of the 1960s and 1970s revealed that scientists were experimenting on people without their knowledge, and the first US governmental bioethics commission was set up in order to recommend limitations on scientists.[22] Therefore, the very institution of public bioethical debate was invented to put limits on scientists' actions, which encourages arguments on a slope: "Above here is acceptable, below here is unacceptable."

Second, since these debates are ultimately input for law and regulation, they conform to the nature of law in liberal democratic societies, which is that the citizens are free to do anything they want as long as there is not a law against it. And, there can only be a law for various good reasons, such as that your action would harm someone else. Therefore, all of these public bioethical debates have to argue against the default mode of US society, which is total freedom, which is located near the bottom of all of the slopes. It is near the bottom because there will always be someone who wants to use their freedom to clone themselves, genetically engineer their child to be taller, practice infanticide, or have a doctor help end their life because they no longer feel like living. That is, the nature of liberal democratic societies requires people to identify the bottom of the slope and to argue against it by taking a position further upslope.

So far with my metaphor I have established a slope with a top, bottom, and continuum in-between, where the top is a meritorious action and the bottom is the nightmare of the people standing at the top. To continue to build this metaphor, the terrain on the slope is the qualities of the act at that slope location on which the locations can be compared. For the first decades of the HGE debate, the terrain was both genetic trait and target. First, on traits. Some genes lead to bodily traits that are neurological disorders like Tay-Sachs. Some genes contribute to the trait of tallness or having blue eyes. Genes have something to do with the trait we call "intelligence," although exactly how that works remains a mystery. These traits are ordered along the slope following our values. For example, a long-standing version of the debate is to order these by the extent they are "diseases." Second, the terrain was also target, and the target is the entity that would be changed. This debate has been constructed since its earliest days as HGE targeting individuals (somatic HGE) in the upper part of the slope or the species (germline HGE) in the lower part of the slope. The point is that the terrain of the slope consisted of combinations of genetically derived traits and targets of the modification, ordered from the most to least acceptable to modify.

More later in the chapter, but for now I would say that in the public bioethical debate about HGE there is very high consensus that at the top of the slope is "somatic gene therapy," which means relieving the genetic diseases in individuals so that those changes are not passed on to their offspring.

The bottom of the slope was created in reaction to the eugenicists, and therefore it is no surprise that the bottom is a dystopian eugenic society. Scholars writing about the slippery slope in the HGE debate who explicitly describe the bottom see it as "morally condemnable eugenics such as the creation of a genetic 'super class' of privileged individuals"; "eugenics, the attempt to create a super race of human beings through the use of genetic technology"; or "eugenic abuse or even the creation of Frankenstein monsters."[23]

There are conservative and liberal versions of the bottom of the slope. In Jonathan Haidt's terms, religious and cultural conservatives value avoiding harm, promoting fairness, loyalty, authority/respect/obedience, and bodily and spiritual purity.[24] They are, for example, in favor of retaining the human species as it currently exists. Fifty years ago this view was more acceptable in public debates about science, and many of the critics of HGE had this basic orientation. Their dystopian bottom of the slope was summarized by the 1931 novel *A Brave New World* by Aldous Huxley, which depicted a society focused on science and efficiency, where emotions and individualism were driven out of people. The people in the Brave New World were bred in eugenic fashion to be in particular social classes. But, the primary feature of the bottom of the slope is that it violates bodily and spiritual purity as technology is ultimately dehumanizing, resulting in people being more object-like—in Huxley's dystopia, humans were bred in bottles. As Leon Kass stated in the early 1970s, "we are witnessing the erosion, perhaps the final erosion, of the idea of man as something splendid or divine, and its replacement with a view that sees man, no less than nature, as simply more raw material for manipulation and homogenization."[25]

For reasons I will discuss in the next chapter, the people who had these concerns largely left the debate, leaving only what Haidt calls liberals, who have a more narrow conception of morality, that is, focused on only avoiding harm and promoting fairness. This has resulted in a different eugenic dystopia at the bottom of the slope, represented by the 1997 movie *Gattaca*. In that dystopian society, children's genetic qualities are selected by their parents in line with a very rigid genetic hierarchy with no mobility between classes, resulting in durable social and economic inequality. Dehumanization is not a theme in this dystopia, but the eugenic society is wrong because it is not fair to those bred into the lower slots in society. The dystopia lies in a particularly powerful mechanism for class inequality that most academics would be opposed to.

Everyone in this debate of whom I am aware has decided to get on the slope with somatic gene therapy, and participants vary widely about where they would ideally stop. There are, of course, modern-day eugenicists in these debates who do not see a downward slope at all and actually hope for the day when the nightmare of the critics comes to pass with children being designed to maximize their intelligence and so on.[26] That said, those academics who yearn for what others see as the bottom of the slope are a minority. Finally, it is important to note that while there is a strong consensus about the top and bottom, there is much less consensus on how the middle of the slope is ordered.

Slippery Slopes in the HGE Debate

To build the next piece of my metaphor, why is the terrain on the slope slippery? As two legal theorists write, small steps down slippery slopes are more likely in the absence of a sharp line between cases, with the cases aligned as a "gradient or continuum," where "arguments or policies are connected by a series of small (perhaps infinitesimally small) steps."[27]

There are two features that encourage slipperiness along the terrain of the slope. One is "continuity vagueness," where the terrain has a continuous measure.[28] For example, if we said that a physician was allowed to facilitate the death of a patient if they only had a 10% chance of surviving the next year, this could easily become 11% because the difference between 10% and 11% is arbitrary. (While durable working compromises on positions subject to continuity vagueness do occur, the conditions for such a compromise do not exist in the HGE debate.)[29]

The second quality of the terrain that leads to slipperiness is "similarity vagueness," where "measurement is not possible, irrelevant, or when it depends, at least in part, on imprecise components." Think of the example of "knife," which in English refers to a huge range of objects with different purposes.[30] A better way to put the similarity vagueness problem for an ethical debate is that the distinction that *can* be made between the steps on the slope does not hold ethical weight, such as knives with wooden handles vs. knives with plastic handles. What is the moral difference between wood and plastic handles?

Until now, continuity vagueness has not been very relevant to the HGE debate because genetic traits and targets have been the terrain, and these are not continuous measures. Rather, the target is categorical (somatic vs. germline) and traits result in different phenotypic experiences that resist commensuration into one scale. For example, the value of height and eye color cannot be quantitatively compared. While continuity vagueness has not yet been important, I will argue that it will be.

On the other hand, similarity vagueness has been a major supplier of grease on the slope of the HGE debate, as what has often driven the small steps down the terrain of the slippery slope are the vague concepts of "health" and "disease." For example, what is the experiential difference between predisposition to heart disease vs. predisposition to breast cancer? The difference of damage to the organ due to cell death vs. runaway cell proliferation does not hold moral weight. If the consensus had been that HGE to treat predisposition to heart disease was acceptable, this position will likely slip to the next spot down the slope when no defensible distinction can be made between heart disease and cancer.

Psychologists see slippery slopes as the result of a basic cognitive process.[31] Public bioethical debates are particularly slippery because of the professional backgrounds of the participants.

A legal theorist writes that judges are particularly prone to slipping down slopes because of their tendency "to place a premium both on drawing non-arbitrary, rationally defensible lines and on maintaining a coherent, consistent body of case law within a particular jurisdiction."[32] If judges have this quality, academics have a double dose, particularly the non-arbitrary, rationally defensible part of the description.

That is, similarity and continuity vagueness are the main drivers of change in bioethical debate because the participants are almost exclusively academics, and academics are rewarded for rational and logical consistency. In this logic, if you approve of position X, and position Y is just like X, then you must approve of Y. The claim that X is the same as Y (i.e., similarity or continuity vagueness) can be justified through casuistry,[33] which is how American law is structured,[34] or through logical deduction from higher level principles, which is how analytic philosophy is structured.[35] The profession of bioethics, the dominant group in public bioethical debate, has a very strong emphasis on the deductive version of logical coherence due to the influence of analytic philosophy. In general, in a debate dominated by academics, and by bioethicists in particular, the slope will be particularly slippery, especially in the case where the debate is input to law and policy.[36]

In a departure from the existing literature on slippery slopes, I think we must consider that the "similarity" that provides the grease begs the question, "Similar by what standard?" The fountain pens of Thomas Jefferson and Barack Obama are undoubtedly quite similar if the standard is "effectiveness of putting ink to paper" but quite dissimilar if the standard is "age." In a debate about ethics, the standard is moral values. Two points on the slope can only be made "the same" through one of these two slippery slope processes if they are justified using the same moral value. Using HGE for sickle cell anemia and for adult-onset heart disease can be made "the same" if both are justified by the value of the relief of suffering.[37]

Whose values? The answer can only be the participants in these debates, but where do those values come from? Undoubtedly one source of values of the participants is the values of the society, and many participants implicitly or explicitly claim they are representing the public's values.[38] In general, and for example, if the country increasingly believes in bodily autonomy, this would be reflected in the HGE debate.

A larger source of values for participants is probably the values used in other public bioethical debates, and there will be professional pressures to adhere to such values. For example, the value of individual freedom and autonomy is used in many bioethical debates, and the use in one undoubtedly increases the odds of being used in another. Moreover, changes in who participates in these debates over time has an effect. For example, the debate was once heavily influenced by theologians who used secular translations of their values. This group has largely left the debate, along with the values they tended to hold. This change in values will mean that particular instances of slipping will or will not occur.

Barriers

The classic slippery slope argument is an argument for the status quo—that we should go no further than where we currently are. But, change is often good, as long as we avoid the bad. As philosopher John Harris wrote, "Slopes are only slippery if they catch us unawares and we have strayed onto them inadequately equipped."[39] The main piece of equipment is limits, and as Granville Williams wrote long ago, "all moral questions involve the drawing of a line."[40] Two legal theorists note that "slippery slopes . . . can sometimes be resisted by standing on easily enforceable bright-line rules." Buchanan and his colleagues, in their influential book on HGE, call these barriers "moral firebreaks."⁴¹ I will conceptualize these limits as barriers on the slope that arrest further slippage.⁴² It is the rise and fall of barriers that shape the debate over time, and the rest of this book will be about these barriers—why certain barriers fell, and whether barriers further down the slope will hold.

There are participants in bioethical debate who accept the idea of barriers on slopes and those who do not. To believe in a barrier, you have to believe that there are acts that we should not do, even if other circumstances change. In philosophy terms, you need to be more of a deontologist. People with this view will use the term "ban" in making recommendations about future research.

A second type of participant does not believe in barriers but rather in speed bumps, which are conditions that must be met for moving further down the slope. In philosophy terms, these people are more consequentialist, and this is the general orientation of most scientists and bioethicists. They want to allow technology to develop right up to the barrier, and then weigh the consequences of action and inaction, and decide whether to move forward. People with this view tend to use the term "moratorium."

The pragmatic reality is, of course, that even if people intend to produce an impenetrable barrier, future generations may knock over that barrier when their values change. Barriers slow down technology's applications so that people can evaluate whether going forward advances our values. Barriers also prevent technological ability driving the debate forward on its own.

Scholars typically argue that their barrier—their limit—is in the most ethically defensible position. In a more pragmatic vein, I will instead argue that scholars should advocate for a barrier that is the most morally defensible position near their actual position. For example, it has been noted that many Christian theologians, were it not for fear of slippery slope processes, would not defend the germline barrier that is roughly built on the value of following nature or God's plan (of which more later). But, in the words of one theologian, in modern society where it is not possible to argue for a purpose for the human, "by default we fall back on the 'natural.'"[43]

The final question is then: What makes a strong barrier? A strong barrier separates two steps on the slope that cannot be made similar through similarity or continuity vagueness. In the example of the somatic/germline barrier, somatic HGE means modifying an existing individual, and germline means modifying the species, and this is a strong barrier because it is difficult to make "individual" and "species" the same. There is no continuum between the two concepts; or, at least in Western ontologies, the two concepts are opposites. Therefore, a strong barrier has the terrain above and below the barrier described as dis-similarly as possible, which limits the possibility of any similarity slippery slope mechanisms. A strong barrier does not have apples above and oranges below, but apples above and ball point pens below.

As an example, Figure 1.1 shows three hypothetical somatic/germline barriers of different strengths. The first frame shows the very strong barrier. Above the slope line I label values, and below the line I label the terrain. The terrain on the different sides of this barrier are not subject to either continuity or similarity vagueness and thus cannot be made "the same." For example, the target upslope of the somatic/germline barrier is "the existing human body" (soma) and the downslope is "the species." I will often write that such a barrier has a good design.

One of the reasons the acts cannot be made the same is that the values used in the debate provide a positive evaluation of the acts on the upslope side of the barrier and a negative evaluation of the acts on the downslope side. If one of the values is non-maleficence (avoiding harm), the acts above the barrier are considered consistent with that value (e.g., they are safe), and below the acts are

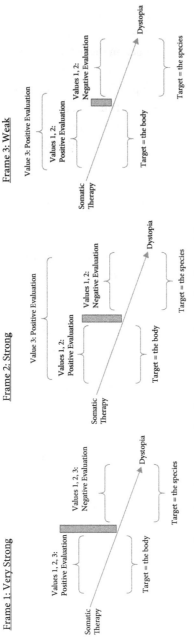

Figure 1.1 Hypothetical somatic/germline barrier strengths on the slippery slope

inconsistent with that value (e.g., they are not safe). Metaphorically, this holds the barrier at that place on the slope.

The value that overlays the act on the terrain defines what is relevant about the act on the terrain. For example, if beneficence (doing good) justifies part of the slope, the acts on the slope are ranked by the amount of good they do. It if is non-maleficence, they are ranked by how safe they are. Critically, barriers become vulnerable to slippery slope processes when the acts downslope get redefined by new values and are then subject to continuity or similarity vagueness with the acts above the barrier.

The second frame in Figure 1.1 shows a weaker, yet still strong barrier. Values 1 and 2 have positive evaluations above and negative evaluations below the slope. But, there is a group in the debate claiming that value 3 not only justifies acts upslope but also justifies acts far below the barrier. This is the more revolutionary approach to moving down slope because it redefines the terrain of a large section of the slope.

For example, starting in the 1980s, the value of "autonomy" came to the fore, which meant that a couple should have the ability to decide what traits their children would have. Some participants in the debate wrote that this value should justify acts not only above the barriers of the era but far below as well because, for example, the value of autonomy could mean that people could decide to have enhanced children.

Such a move is difficult to accomplish—indeed, while it threatens, it has not been done in the history of the debate. The reason this is difficult to achieve is that it is radical—it requires asserting that a huge range of acts previously thought to be morally wrong are actually acceptable, and that those who support values 1 and 2 are not slightly wrong but totally wrong for holding the barrier in its current place. Such a radical claim would be: Up until now the current barrier has meant that creating a child with enhanced intelligence was far below the barrier near the bottom of the slope, but we are arguing that the reproductive autonomy of the parents means it should be acceptable. The reason that public bioethical debate does not like radical change is that it is at the interface between scientists and society, with scientists being very concerned about how the public reacts to their work. Bomb throwers are not appreciated. I will show attempts like this in the history of the debate—all of which failed.

Frame 3 shows a weaker barrier that is analogous to Frame 2 except that value 3 reaches only slightly over the barrier to redefine acts that are relatively close to the acceptable acts upslope of the barrier. This subtly redefines the acts just below the barrier to be the same as those above, resulting in continuity and/or similarity vagueness. The acts on both sides of the barrier become "the same," and the barrier falls.

Compared to the situation depicted in Frame 2, Frame 3 is not radical. In extending the positive evaluation of value 3 barely below the barrier, we may find that the supporters of values 1 and 2 may not disagree with the acts that are newly justified—they were just supporting the barrier in its location to avoid the really offensive acts far below. But, the effect is the same—if the acts on both sides of the barrier are supported by the same value, and are then the same, the barrier is felled. Without a barrier, eventually the justified acts in the debate will slip down to the region of the slope that was most concerning. Empirically, I consider a barrier to have fallen when a majority of the participants in the debate agree that it has.

We now have all of our tools in hand for understanding the history and future of this debate. I start my narrative in the next chapter with an examination of the first two barriers erected on the slope of this debate in the early 1970s, and how, beginning in the 1980s these barriers were weakened. In Chapter 3, I examine the debate in the CRISPR era and focus on how an influential American pseudo-governmental commission advocated removing the germline barrier, yet provided the intellectual justification for putting a new, strong barrier lower on the slope. In Chapter 4, I turn to a more speculative direction, identifying barriers further downslope that have either been proposed in the literature or which would have good design characteristics. In Chapter 5, I conclude by recapping my argument, identifying some general lessons about slippery slopes in public bioethical debate, and describing a future debate where participants acknowledge slippery slope processes.

2
The First Barriers in the Human Genetic Engineering Debate

I now turn to the history of public bioethical debate about HGE from the 1950s forward.[1] Through this history we can see the processes that build and collapse barriers, and this knowledge will allow us to better predict which future barriers will and will not hold.

Obviously there are many ways the genes in a society are indirectly affected. Migration patterns, fertility rates, social norms about who should marry who, and much else influences the collective human genome. The HGE debate began in the eugenics debate, which concerned the intentional and directed human control of such processes. Starting in the late 19th century, the goal of eugenics was to give "the more suitable races or strains of blood a better chance of prevailing speedily over the less suitable."[2] The stated motivation for eugenics was that, due to the conveniences of modern society like medicine, genetically weak individuals were not culled by evolutionary forces but were allowed to live to reproductive age. What was worse, it was thought that the weakest of these people—the poor in the minds of the British elites concerned with eugenics—were reproducing even more than the genetically superior people, degrading the human gene pool. Francis Galton was the inventor of the term eugenics, and he wrote that the "survival of the fittest" was no longer operative due to modern conveniences such as "the minimum wage, the eight-hour day [and] free medical advice." One of Galton's disciples would conclude that natural selection had been replaced by "reproductive selection," which gave the advantage "to the most fertile, not the most fit."[3]

Most important for the later debate, the eugenicists had a radically expansive notion of which traits were genetically inherited. They thought if you looked at the leaders of society such as intellectuals, artists, and scientists, you could see the result of good breeding. The bad genetic traits included criminality, alcoholism, sexual immorality, and "feeblemindedness."

Up until the early 20th century the eugenicists assumed that superior and inferior traits were clustered in races, ethnicities, and, particularly in Britain, classes. All sorts of schemes were developed to encourage people with what were thought to be superior genes to have more children and to encourage or force those with inferior genes to have fewer. Eugenicists wanted to improve the humans living in a nation or, depending on your outlook, stop genetic quality from degrading further.

In the American context the eugenicists had their greatest success in 1924 with the immigration control act that limited immigration from nations thought to be the source of "bad" genes.[4] The bill was signed by President Coolidge, who earlier had said that "America must be kept American. Biological laws show . . . that Nordics deteriorate when mixed with other races."[5] Another infamous indicator of the eugenics mindset occurred when US Supreme Court Justice Oliver Wendell Homes justified the forced sterilization of a 17-year-old "moral imbecile" by claiming that "three generations of imbeciles are enough."[6]

Eugenics was not necessarily a right-wing political movement, but at least in the United States and the United Kingdom it was taken up by social progressives trying to resolve social problems. For example, American liberal and mainline Protestant denominations, as well as liberal Jews, were strong supporters of eugenics as a means to uplift the poor.[7] Of course these motivations were not pure by our contemporary standards—there was strong race and class bias in progressive support for eugenics before World War II.

By the 1930s the connection between good and bad genes and class, race and ethnicity began to be questioned, with many pointing out that these eugenic ideas were thinly disguised class and race prejudice. One scientist at the time wrote that eugenics had become "hopelessly perverted" into a pseudoscientific facade for "advocates of race and class prejudice, defenders of vested interests of church and state, Fascists, Hitlerites, and reactionaries generally."[8] It was the Nazis who took race-based eugenics to its apotheosis with genocide against "inferior" races. In the words of the foremost historian of the eugenics movement, after the eugenic policies of the Nazis were fully exposed, people realized that "a river of blood would eventually run from the [German] sterilization law of 1933 to Auschwitz and Buchenwald." Thus, the mainstream eugenics movement was dead.[9]

However, the Holocaust was only the end of coercive race- or ethnicity-based eugenics. Eugenicists continued to argue that humans were extremely unequal in their genetic contribution to the human gene pool; it is just that these inequalities are not structured by race, ethnicity, or class. They argued that eugenicists should identify valuable genes found in any group and encourage those with these genes to reproduce. What the eugenicists of the 1950s had in mind were voluntaristic, race-neutral schemes to get individuals to breed with the right people and for the right people to have more babies.

An embryonic form of a later barrier began to emerge with the specific concern of eugenicists with the "genetic load." The genetic load was the sum of the disease-causing genes thought to be spreading rapidly in the population, which was the result of modern medicine keeping people alive who should, from a Darwinian perspective, die before reproducing. According to one influential scientist, given enough time, "the germ cells of what were once human beings would be a lot of hopeless, utterly diverse genetic monstrosities."[10]

But, like the pre–World War II eugenicists, these "reform" eugenicists were not only concerned with stopping the spread of "disease genes" in the population but also with encouraging the spread of genes that represented the "ideal man." For example, British scientist Sir Julian Huxley wrote that he wanted to use eugenics to increase "man's desirable genetic capacities for intelligence and imagination, empathy and co-operation, and a sense of discipline and duty."[11] That is, the distinction between eugenics to reduce genetic disease vs. eugenics to enhance the human population began to form in this early debate, even though most eugenicists were in favor of both. This distinction would later become a barrier. It is important to remember that in this era the target for modification was the species not the individual.

The eugenics debate birthed the modern HGE debate in 1953 when Crick and Watson published their paper describing DNA, which further detailed the chemical basis of heredity. By the 1960s, people in the eugenics debate began to realize that if genes are chemicals, then the genes of people could be changed chemically, avoiding the need to influence people's mating decisions. Moreover, with mating, the baby's qualities were limited to the genes the parents had. But if genes could be modified, it was possible to imagine inventing new "super" traits. We would "have the potential to create new genes and new qualities yet undreamed" of in the human species.[12]

However, by the mid-1960s the eugenicists became intimidated by the task of setting the proper goal for human evolutionary progress. They had to, using their terminology, define the "ideal man," but they lost their confidence that they alone could make such a definition. As a scientist at one of the reform eugenicist conferences noted in the mid-1960s, how the new technology would be used "will not be decided by scientists alone," but rather by "enlightened and broadly based public opinion."[13]

The public *did* start paying attention. Popular books began to be written on the subject, and other academics began to take notice of the debate. Most notable were the theologians, who saw the question of what the "ideal man" should be as impinging on theological turf. For example, theologian Paul Ramsey accused scientists of advocating a "surrogate theology" of the "cult" of "messianic positivism" where scientists decide what our values should be. Ramsey also accused one of the influential scientists of the era as having goals that were simply "an extrapolation of what we should do from what we can do."[14] More generally, a debate ensued about what the goals of HGE should be. What would the "ideal man" be like, and who would decide?

At the same time, the public was beginning to question the wisdom of scientists. According to sociologist Kelly Moore, "by the middle of the 1960's, scientists were being blamed for, among other things, the war in Vietnam, alienation ... and a multitude of environmental problems."[15] In another scholar's summary, "public confidence in technological development as the key to social progress gave way to disenchantment."[16]

A statement summarizing a 1972 conference makes it clear that restricting the debate to scientists was ending, and that a slope with barriers to define limits on scientists' actions was going to be built. The back cover of the conference proceedings stated: "The scientist often works in what amounts to a vacuum of public opinion, despite the fact that he wields potentially more power over our destiny than all the Presidents and Premiers combined. . . . Society runs a grave risk by granting carte-blanche to the scientific community, particularly in the area of genetics."[17] What would those limits—those barriers—ultimately be?

The First Barriers

A particularly powerful reminder that the target of the eugenicists' actions was the species (not the individual), even for health, is H. Bentley Glass's presidential address to the American Academy for the Advancement of Science in 1970, later published in *Science*. He thought that "the nature and personality of man must change," and thus "no parents will in that future time have a right to burden society with a malformed or a mentally incompetent child." Moreover, "if every couple were permitted to have only two children, or to exceed that number only upon special evidence that the first two are physically and mentally sound, a mild eugenic practice would be introduced that is probably all mankind is prepared to accept at this time."[18] Again, the eugenicists wanted to stop individuals from having genetic diseases, but primarily because diseased individuals would contaminate the human species' gene pool, not to relieve the suffering of any one individual.

This early HGE debate was only about the germline, and participants focused on both diseases and enhancements. For example, Glass wrote that he was looking to create the "good man," who would have "freedom from gross physical or mental defects, sound health, high intelligence, general adaptability, integrity of character, and nobility of spirit."[19]

All this talk of creating the perfect human through genetic engineering generated social controversy, and basic genetic science was threatened by the bad publicity from the eugenicists who wanted to perfect the species. This threat was described in 1970 by Bernard Davis who wrote in *Science* that the "exaggeration of the dangers from genetics will inevitably contribute to an already distorted public view, which increasingly blames science for our problems and ignores its contributions to our welfare. Indeed, irresponsible hyperbole on the genetic issue has already influenced the funding of research." Citing some eugenicists of the era, who have "caused wide public concern," he also noted that discussions about HGE "have tended toward exuberant, Promethean predications of unlimited control and have led the public to expect the blueprinting of human personalities."[20] Here we can see an early proposal for the ordering of the slope, with an enhancement like "blueprinting of human personalities" at the bottom.

The scientists created barriers that placed all the socially controversial applications scientists did not want to do downslope and the applications that they wanted to conduct upslope. Scientists simultaneously created and then retreated behind two barriers: the somatic/germline barrier, and the disease/enhancement barrier. Again, somatic means that the genetic modification would only impact the genes of an existing human and not be passed on to offspring. Germline means the edited genes are passed on to descendants and ultimately the species. "Disease" was not considered problematic to define, for reasons I will get to shortly, and roughly coincided with all the traits of the "genetic load" that eugenicists thought was bearing down on humanity. "Enhancements" came to mean all those traits of the eugenicists' dreams like intelligence, artistic ability, and good character.

The creation of the first barrier came as scientists like Davis implicitly redefined the value that would justify HGE as beneficence (relieving suffering) and not "perfecting humanity," which created a barrier between HGE for "diseases" and those for "enhancements." This barrier would not only calm the public, but the upslope applications were seemingly going to be possible soon. By this point, scientists had been able to synthesize genes and introduce them to bacteria, and Davis was hopeful that scientists would soon synthesize and modify human genes in a test tube.[21]

This barrier was also possible because at the time the known genetic diseases were monogenic, the result of an error in one gene, and changing one gene seemed plausible. On the other hand, enhancements like intelligence, assuming it could even be defined, would be polygenic, based on multiple genes interacting in a way that was not understood. Davis argued for the barrier when he wrote that most geneticists had "more restrained second thoughts" about the possibility of engineering polygenetic behavioral traits. While Davis advocated for the disease/enhancement barrier, he did not advocate for a somatic/germline barrier and, indeed, thought that germline modification of humans for these monogenic diseases was more efficient than somatic changes. With somatic you would have to modify all people born with the disease, but with germline nobody would be born with the disease.[22] Others would argue for the somatic/germline barrier.

A few years after Davis's article, scientists Theodore Friedmann and Richard Roblin described the somatic editing they wanted to do not as human genetic engineering but "human gene therapy." The term "therapy" connected these actions to the value of beneficence—the relief of suffering—that has traditionally been the province of medical science and thus to fighting disease. Their article, titled "Gene Therapy for Human Genetic Disease?," pointed to the acts of editing in their acceptable part of the slope by beginning with the statement "at least 1,500 distinguishable human diseases are already known to be genetically determined."[23] Pointedly, they did not talk about the genetic enhancements of the eugenicists' dreams. Strategically, it makes sense to build this barrier as it allows for the scientific community to engage in somatic HGE for monogenic disease that was potentially imminently possible and distinguish these applications from those that would be many decades away. This would keep what they wanted to do on the good side of public opinion.

Friedmann and Roblin cite one of the first articles to mention "gene therapy," and that article reveals the fear of controversy that motivated the genetics community to build barriers. In that article, after a scientific examination of the possibility of genetic modification using viruses as delivery vehicles for the new DNA into the human body, the author writes that "the rapid development of molecular biology and its potential applications to medicine have been likened by some people to the rapid development of theoretical physics and its ensuing application to atomic technology." The physicists' work resulted in destructive weapons, and "rightly or wrongly, scientists have been blamed for these technological creations," and therefore "many already fear genetic research." He concludes that "this fear of genetic research might be relieved to some extent if the term 'genetic engineering' were abandoned in favor of the term 'gene therapy.' The term 'genetic engineering' has the revolting connotation to many of impersonal scientific manipulation of the future of human life and offends the dignity of many."[24] This scientist, like others, wanted to distinguish between therapy and enhancement, as well as "the future of human life"—the germline—from the treatment of individuals. These statements and others like them slowly built the somatic/germline and disease/enhancement barriers.

It is not easy to say which of these barriers was upslope of the other because different participants had different rankings of what was acceptable and unacceptable. The best way to consider the use of these barriers using my metaphor is that they were at the same place on the slope. Not everyone used both barriers.

Figure 2.1 shows the general terminology used in the debate for the combinations of the two barriers. Cell one of the figure

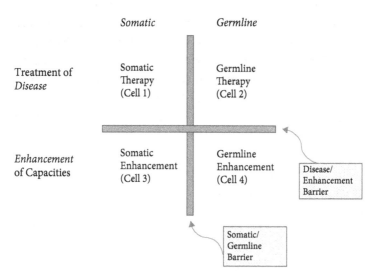

Figure 2.1 Combinations of barriers, early HGE debate

represents the case where the participant argued for both the somatic/germline and the disease/enhancement barriers, and therefore only advocated what was above both: "somatic therapy." Cell two represents those who believed in the disease/enhancement barrier but not the somatic/germline barrier. They would argue for "gene therapy," be it of the soma or the germline. Cell three represents those who believed in the somatic/germline barrier but not the disease/enhancement barrier, which would mean advocating somatic enhancement. This combination was rarely or never advocated. Cell four represents those who did not believe in either barrier, and therefore were in favor of therapy or enhancement of the soma or the germline. Cell four was where participants in the debate placed the now discredited eugenicists who desired the physical and behavioral perfection of the species. By the 1980s the debate reached consensus on these barriers, which remained the only barriers on the slope for many years.[25]

A debate at a 1971 conference, while the barriers were still forming, is illustrative of how consensus was built that these would be the barriers that would structure the debate. In the debate between theologian Paul Ramsey and scientist W. French Anderson, Anderson tried to support the disease/enhancement barrier but rejected the germline barrier, which would have collapsed the distinction between cells one and two in Figure 2.1. He defined all the terrain on the slope above the disease/enhancement barrier as "gene therapy"—the "treatment of hereditary diseases"—which suggested that doctors should heal the species (the germline).[26] This would make the value that justified HGE as only beneficence (relief of suffering), make the terrain only "trait," and implicitly bring back the arguments of the reform eugenicists who were concerned with improving the genetic health of the human species.

Ramsey pushed back by saying the word "therapy" can only be applied to an existing body, not a species, which implies the somatic/germline barrier. He proposed a term to replace "gene therapy"— "genetic surgery"—which would have even more explicitly defined the target of the modification as the existing individual. That is, in English you could stretch the meaning of words and say you are going to provide "therapy" to people's offspring by modifying them, but it is nonsensical to say you can conduct surgery on the human species or on a human embryo. More importantly, "surgery" also connects these acts more clearly with medicine, further reinforcing the disease/enhancement barrier.

Ramsey assigns all of the terrain above the two barriers to medicine, essentially saying that scientists should commence trying to figure out how to use gene therapy to treat individuals who have diseases.[27] That is, the fiercest critic of HGE had defined a space upslope of the two barriers that was acceptable to him, and this barrier came to be accepted in the debate, with any germline modifications and any enhancements downslope. While his "genetic surgery" term never caught on, Ramsey had stepped onto the slope, accepting somatic gene therapy, presumably confident that the two barriers would hold. Most others joined him. Proponents of somatic gene therapy came to realize that at minimum the somatic/germline barrier was useful for limiting controversy, given that scientists had no idea how germline HGE could be accomplished.

Why These Barriers Were Strong

Design Strength of the Disease/Enhancement Barrier

From the early 1970s through the 1980s the two barriers were extremely strong. Let us begin with the disease/enhancement barrier, and with the terrain on both sides of the barrier, which was the trait. Could the traits on the two sides of the barrier be made similar, thus toppling it? The very limited knowledge of human genetics in the 1970s and 1980s meant that, unlike later years, "disease" (e.g., cystic fibrosis) and "enhancement" (e.g., intelligence) on the two sides of the barrier were radically dis-similar. This was before the human genome project, before genetic tests, before any thought of sequencing someone's entire genome, before the discovery of epigenetics—before most of the knowledge we take for granted today. "Disease" traits were *only* those clearly observable in the body of a human, and the trait had to be clearly different from normal human variation. Genetic diseases had to be obvious and severe and therefore typically observed in children. For example, Friedmann and Roblin started their 1972 article by mentioning phenylketonuria and cystic fibrosis—two extremely serious Mendelian recessive diseases that become apparent shortly after birth.[28]

Lacking genetic tests, the only way that a disease was defined as genetic was that it was so clearly distinct from normal variation that your parents or their relatives were known by the family to have displayed similar symptoms. Today we might think of genes that increase the odds of getting cancer at age 50, but nobody back then could look at a family tree and know that this trait was inherited because there are all sorts of reasons people get cancer. In sum, the bodily experience of what were then genetic diseases and the enhancements of the eugenicists' dreams were so vastly different there was no risk of similarity vagueness between the traits on the two sides of the barrier.

Moreover, the scientists who helped construct the barrier in the 1970s made it even stronger by describing the upslope genetic diseases as monogenic and the enhancements downslope as polygenetic, seemingly grounding the barrier in the nature of human biology. By this time a large number of diseases were known to be the result of one errant gene passed down following the pattern of Mendelian inheritance. This could be usefully distinguished from polygenic enhancements. As Davis would write in 1970, "the most interesting human traits—relating to intelligence, temperament, and physical structure—are highly polygenic." Yet, he continued, "the study of polygenic inheritance is still primitive."[29] Similarly, in 1980, gene therapy researcher W. French Anderson and bioethicist

John Fletcher wrote, when debating in the *New England Journal of Medicine* when somatic gene therapy trials should begin, that "we are currently attempting to influence in a known way only single-gene defects. Intelligence, personality, fertility, organ structure, and physical, mental and emotional characteristics are all presumably controlled or influenced by vast arrays of genes about which we know little or nothing."[30]

Conceptualizing "disease" not only by the characteristics of the trait but also by the number of genes involved does define the terrain as the "number of genes," and thus in principle makes the barrier vulnerable to a continuity vagueness slippery slope mechanism. However, the fact that these numbers of genes were portrayed as one vs. more than one gave this barrier a categorical instead of numeric definition.

That is, if the barrier had been built at the space between traits caused by 16 vs. 17 genes, it would have been very susceptible to continuity vagueness. However, there is a strong cognitive difference between zero and one, as well as one and more than one. At least Westerners make major linguistic distinctions between these two, with additional terms covering differences on the number line. Zero is also "none," "without," "nothing," and so on. "One" is also "a," two or three becomes "few," and more than three or four is "many." At least in English, there are words to distinguish between one and more than one, but there is no word that can be used to distinguish between five and six, six and seven, and so on. So, at least Westerners are primed to recognize a distinction between one and more than one.

Finally, the barrier was further reinforced in the early 1980s because discussion focused not only on monogenic diseases but also on the most extreme monogenic diseases. The reason is that in the 1980s much of the debate concerned when to start clinical trials for somatic gene therapy (Figure 2.1, cell one), and the only "first in a human" clinical trials that could be approved would be one with an extreme risk-benefit ratio.[31] That is, given the extremely high risk in these first human trials, the only justifiable trial would be one where the quality of life was so severely impaired that any benefit was worth the high risk.[32] Indeed, the first approved clinical trial for an inherited disease was for adenosine deaminase (ADA) deficiency, which leaves people without a functioning immune system.[33] Thus, the traits being discussed on both sides of the barrier became even more distinct. In later years knowledge would fill in between these extremes, weakening this barrier, but in the early years the barrier was strong.

Values Supporting the Disease/Enhancement Barrier

Again, a barrier is strong if all the values that support it have a positive evaluation of the actions upslope and a negative evaluation of the actions downslope. For the disease/enhancement barrier, the radically different traits on the two sides made the values clear. For the value of beneficence, promoted by all in the debate, above the barrier the acts on the terrain were interpreted as the beneficent relief of suffering, below the barrier interpreted as not beneficence, as something more like egotism by wanting to be better than everyone else.

In this era, the connection of the value of beneficence and the disease traits upslope was reinforced because this connection between beneficence and human disease—defined as bodily dysfunction—was synonymous with the well-respected medical profession. Put simply, this is what medicine "was" in this era, so relieving the suffering caused by a trait like sickle cell anemia made perfect sense. To foreshadow later pages in this book, in future years the profession of medicine would not only be associated with healing suffering from bodily dysfunction but also with treating people's desires about their bodies with practices like plastic surgery in what scholars call "wish-fulfilling medicine."[34] This would later weaken the connection between beneficence and genetic traits.

Also supporting the disease/enhancement barrier was the value of following God's will—a value with some constituency at this point, whether explicitly or implicitly stated. Above the barrier was, in Christian terms, repairing diseases that were the result of the Fall of man. Below the barrier was not following God's will but rather trying to improve on God's design for the species.

Upslope were monogenic traits that could conceivably be understood and thus safely manipulated. Downslope were polygenic traits that were utterly mysterious at the time so modifying them was quite dangerous. Therefore, another consensually held value—non-maleficence (avoiding harm)—supported the barrier.

Relatedly and finally, also supporting the barrier was the value of humility and its antonym hubris. Above the barrier with diseases was humility in that all agreed that these diseases were bad and needed amelioration. Below the barrier, in the hubris territory, was thinking that we could define what the perfect human was for all people across all times, as well as the undemocratic arrogance of scientists assuming they knew what positive traits to engineer into the species. While in an earlier time the reform eugenicists would have promoted a value called something like "species progress," which would justify acts of enhancement below the barrier, there were few if any willing to support that value.

Design Strength of the Somatic/Germline Barrier

Turning to the somatic/germline barrier, it appeared even stronger than the disease/enhancement barrier because the terrain was the target of the genetic modification. The target on the upslope side was *the individual* who you could meet and see in the clinic—they could be a relative of yours or written about in the newspaper. Of ethical relevance is that researchers could ask the individual if they wanted to be modified. They existed in the present. On the downslope side the target was *the human species*, who could not be met but was an abstraction. You also could not receive consent from the species, and any change in the species was in the future.

These differences in how the targets of HGE were portrayed make this barrier seemingly immune to slippery slope processes, as the two sides cannot be made similar through continuity or similarity vagueness. We can see why critics like Ramsey thought it was safe to walk onto the slope and advocate for somatic HGE.

Values Supporting the Somatic/Germline Barrier

The barrier was strong because the dominant values in the debate provided a positive evaluation of the acts upslope and a negative evaluation of the acts downslope. As with the disease/enhancement barrier, the value of non-maleficence supported the upslope somatic applications, which had the least odds of harm. If the limitation in our ability or knowledge produced a mistake in somatic HGE, the effects of that mistake would end with the death of the modified person. Below the barrier, with germline, any mistake could spread to descendants and ultimately the population.

Humility and its corresponding antonym hubris also supported the barrier, as the acts above the barrier were cautious and limited, whereas below the barrier it was hubristic to think that mere humans could redesign the species. The value of social equality also supported the barrier, with above the barrier considering people as individuals and below the barrier deciding which traits held by which groups should be the future of the species.

More importantly, above the barrier was the value of respecting "nature" or "God's will," with both concepts representing what humans had not fundamentally intervened in. If modifications were restricted to somatic, above the barrier, nature or God's creation was not being redesigned, humans were simply modifying existing nature as they have always done via medicine and technology.

This value is best seen in contrast with the values used by the reform eugenicists who did not want any barriers. In 1957 reform eugenicist Sir Julian Huxley argued that humans had to control the genetic qualities of the species:

> It is as if man had become suddenly appointed managing director of the biggest business of all, the business of evolution—appointed without being asked if he wanted it, and without proper warning and preparation. What is more, he can't refuse the job. Whether he wants to or not, whether he is conscious of what he is doing or not, he is in point of fact determining the future direction of evolution on this earth. That is his inescapable destiny, and the sooner he realizes it and starts believing in it, the better for all concerned.[35]

Similarly, geneticist Theodosius Dobzhansky wrote in 1962 that "man and man alone knows that the world evolves and that he evolves with it," and "the hope lies in the possibility that changes resulting from knowledge may also be directed by knowledge. Evolution need no longer be a destiny imposed from without; it may conceivably be controlled by man, in accordance with his wisdom and values."[36] Using this value would justify acts far below the germline barrier.

In contrast, responding to another eugenicist of the time, theologian Paul Ramsey would say that the germline HGE proposals of the eugenicists were an improper attempt to elevate human desire over God's will through control of the species:

> [T]aken as a whole, the proposals of the revolutionary biologists, the anatomy of their basic thought-forms, the ultimate context for acting on these proposals provides a propitious place for learning the meaning of "playing God"—in contrast to being men on earth.... [The scientists have] "a distinctive attitude toward the world," "a program for utterly transforming it," an "unshakable,"

nay even a "fanatical," confidence in a "worldview," a "faith" no less than a "program" for the reconstruction of mankind.[37]

This value of following God's will was one of the reasons why the theologians of this period, who were mostly Christians of various types, generally supported this barrier. Modern Christianity has generally held that while we humans are supposed to intervene in the natural world in order to fix its problems, we were to do so by following God's will for nature.[38] Deciding between what was human responsibility and what was God's responsibility has never been easy, but the "reconstruction of mankind" would be in God's domain. The germline barrier was a convenient compromise or generalization for the Christian participants in these debates.[39] Despite the value of human control over the nature of humanity being considered a good in the eugenic era, by the time that the barriers were solidified there were few if any participants willing to defend the values of eugenicists, and the values of the theologians were much stronger.

The Two Barriers in Use

These two barriers functioned as designed, and HGE on targets and traits above the combination of the two barriers—somatic gene therapy—rapidly became uncontroversial. One prominent researcher would write in 1984 that "essentially all observers have stated that they believe that it would be ethical to insert genetic material into a human being for the sole purpose of medically correcting a severe genetic defect in that patient—that is, somatic cell gene therapy. Attempts to correct germ cells . . . or to enhance or improve a 'normal' person by gene manipulation do not have societal acceptance at this time."[40] Similarly, Theodore Friedmann would write in the late 1980s that "few discussions of gene therapy at scientific meetings and in publications still argue its need or potential place in medicine or is ethical acceptability, but rather they emphasize technical questions of efficiency of gene delivery and targeting and selection of suitable disease models."[41] Bioethicist LeRoy Walters examined all available policy statements on HGE across the globe written between 1980 and 1990, primarily written by governments and religious groups, and concluded that "without exception, all 20 of these policy statements accept the moral legitimacy of somatic cell gene therapy for the cure of disease."[42]

The debate then focused on the technical question of whether somatic gene therapy could be done safely. Prominent review articles in this era had titles such as "prospects for gene therapy," "progress toward gene therapy," "the slow road to gene therapy."[43] In 1985, the US government entity that regulated HGE concluded that somatic gene therapy trials could begin once safety and other considerations had been met. Only "diseases" could meet the risk-benefit standard, given that the benefit to healing disease was clear compared to the benefit from enhancing someone. Moreover, the regulator wrote that it would "not at present entertain proposals for germline alterations."[44]

In 1990 the first approved somatic gene therapy attempt for a genetic disease was made on a patient with adenosine deaminase deficiency (as mentioned, a genetic disorder that leaves its victims without a defense against infections). The white blood cells were removed, the healthy version of the gene was inserted into the cells, and the corrected cells put back.[45] Again, this did not impact the reproductive cells of the patient, so it was a case of somatic, not germline modification.

The acts below the two barriers on the slope remained unethical. That would change. At about the same time that government policy made somatic gene therapy totally acceptable, some scholars were advocating removing the barriers that had recently been built on the slope.

A Change in the Participants and Their Values

In the 1980s the barriers were weakened due to a major shift in the participants in public bioethical debate, which brought a change in values. As I have written extensively elsewhere, the "consumer," so to speak, of this debate before the late 1970s was academics and the public.[46] The scholars would present at scholarly meetings, write for scholarly journals, but also try to appeal to the public. This debate was largely about the values, goals, or ends that we as a society should pursue through human genetics. Is our goal to make humanity more intelligent? Is our goal to make humanity healthier? More caring? More artistic? Is it hubristic to think we can design humanity?

In 1970 scientists still had control over what is now called public bioethical debate. The scientists had become very concerned about their newfound powers over nature and concluded that the issues were too monumental to have a small group of scientists decide what to do. For example, Salvador Luria—who won the Nobel Prize in 1969 for discoveries in genetics—said at a conference in 1965 that to "claim the right to decide alone" would be "to advocate technocracy," and that the United Nations and the US National Academy of Sciences should "establish committees on the genetic direction of human heredity."[47] As discussed, some non-scientists like theologians and philosophers were invited to the debate during those early years.

The public was actually paying attention to this emerging technology (and many others at the time) and took these concerns to their elected officials. The elected officials responded in 1973 by creating the first US federal public bioethics commission that would act as an ethical eye on scientists. By and large these commissions did not see their role as participating in public debates, but rather as clarifying the ethical issues for policymakers, particularly in government agencies that had to regulate science. This established that the consumer of ethical discourse about bioethical issues was the administrative state—such as the bureaucracy at the National Institutes of Health. These were the entities that were going to set the policies that would protect the citizens from any dangers from new biomedical technologies.

What values would one of these government commissions use? Since a government commission was in principle representing the public, it could not use the values of the individuals who happened to be commissioners but needed to use the values of the public. Moreover, a commission did not want to have a continuous debate about what those values of the public were for each separate issue that came their way. The first commission established the idea there were three values that were consensually held by Americans that should be applied to all biomedical issues, with this list quickly being modified to four in the debate: autonomy, beneficence, non-maleficence, and justice.[48] In plain English: autonomy means people deciding for themselves, beneficence means doing good, non-maleficence means avoiding harm, and justice means treating people fairly.

With this new approach to ethics came the birth of the new profession called "bioethics" that would specialize in ethics based on these purportedly consensual values. This profession's version of ethics was in contrast to professions like theology that used a different set of values that could *not* be portrayed as consensually held by the public.

The new bioethics profession had an advantage over others in these debates because it was using a set of values preferred by the administrative state. These values were the sort of base minimum that few if any citizens would disagree with and were purported by bioethicists to be universally held. A universally held value would be more comfortable for an employee of the administrative agency because it would be easier to tell the public your ethical decision about HGE was based on the universal values of the citizens and not based upon your own values. This would certainly be easier than basing a decision on values promoted by a theologian whose values cannot be portrayed as consensual.

As the 1970s turned to the 1980s, these four values came to increasingly dominate public bioethical debates on all issues, and values like following nature or God's will declined. Over time, members of professions like theology who would not keep themselves limited to these four values left or did not join the debate while the number of bioethicists grew. But, could these values of the bioethicists still hold up the barriers that had been built on the HGE slippery slope?

Weakening the Somatic/Germline Barrier

One way to dismantle a barrier is by arguing that a positive value actually applies to acts not only above but far below the barrier as well. This is portrayed in frame 2 of Figure 1.1. In the HGE case, the increasingly influential values of the bioethics profession originally made a positive evaluation of the acts upslope of the somatic/germline and disease/enhancement barriers. But, in the 1980s, many influential debaters claimed that the positive evaluation of the terrain using those values upslope of the barrier also applied to the terrain far downslope of the germline barrier. This made HGE applications to the individual and the species subject to similarity vagueness, weakening the barrier.

The Emphasis on Beneficence Undermines the Germline Barrier

The applications upslope of the somatic/germline and disease/enhancement barriers had become totally normalized medical research supported by the values used by bioethicists: autonomy, beneficence, non-maleficence, and justice. In particular, these gene therapy pioneers justifiably focused on the extreme suffering of people who had these extreme genetic diseases. Indeed, it was relieving this suffering that most likely motivated their entire careers, so they were particularly focused on this value. In sum, the first step on the slope, taken years previously, had led to a laser-like focus on the value of beneficence. Influential participants in the debate began to treat beneficence like it was the predominant value, and this resulted in other values that held up the barrier being ignored.

The first somatic therapy trial had not even begun, and bioethicists and scientists were arguing that the germline barrier should be taken down because the only relevant value was beneficence and acts of HGE supported by this value were on both sides of the barrier. In general, those who advocated taking down the germline barrier argued strongly for keeping the remaining disease/enhancement barrier as the bulwark against the slippery slope. As Fletcher and Anderson wrote:

> [S]earches for cure and prevention of genetic disorders by germ-line therapy arise from principles of beneficence and nonmaleficence, which create imperatives to relieve and prevent basic causes of human suffering. It follows from this ethical imperative that society ought not to draw a moral line between intentional germ-line therapy and somatic cell therapy. . . . In our view, a moral line should be drawn between both of these modes of gene therapy—directed towards disorders with the greatest magnitude of suffering, pain, and early death—and efforts at "enhancement" by either mode of therapy.[49]

Some statements, like that above, presumed that beneficence would apply to the species by reducing the odds that a human would be born with a genetic disease. The more serious threat to the barrier was that authors changed the target by identifying an individual human in need of beneficence on the terrain downslope of the barrier where previously only the species had been.

The terrain at the time of the construction of the somatic/germline barrier had on one side the adult existing human, and on the other, the species. The fact that the species could only be influenced by first creating an embryo that becomes a baby was not focused on—after all, eugenicists were not very concerned with individuals. Individual vs. species was a huge difference, and perhaps impossible to make similar.

The first individual identified on the slope below the barrier was the unmodified zygote that would, when a born human, need to be somatically modified to treat their disease. And, it was much more efficient to treat that individual when they were a zygote than when they were a full-fledged human. The fact that this modified zygote would eventually change the species in a small way was not focused upon. For example, LeRoy Walters, the bioethicist who was the chair of the committee at the National Institutes of Health that regulated trials of HGE, extrapolated from relieving the suffering of disease for the individual on the somatic side of the barrier to an individual on the germline side:

> [T]he principal rationale for germline human gene therapy, when it becomes a technical possibility, will be a simple argument from efficiency. . . . Affected offspring could presumably be treated by means of somatic-cell gene therapy in each succeeding generation, but some phenotypically cured patients would probably consider it more efficient to prevent the transmission of specific malfunctioning genes to their offspring, if the option were available.[50]

Reminding us that somatic therapy only works if the scientist can access the malfunctioning cells in the body, Walters continued, "a second rationale for the germline approach is that some genetic diseases may be treatable only by this method. For example . . . the brain cells involved in hereditary central nervous system disorders may be inaccessible to somatic-cell gene therapy."[51] Somatic gene therapy pioneer Theodore Friedmann concurred, also emphasizing that some genetic diseases cause permanent damage in newborns before they could be treated. He wrote in 1989, "it seems unwise and premature to take such a severe position," against germline HGE, "and it has been suggested that the need for efficient disease control or the need to prevent damage early in development or in inaccessible cells may eventually justify germ line therapy."[52]

While these new arguments might have briefly acknowledged that germline HGE would change the species, their overwhelming focus was that a germline change was an unintended side effect of the primary goal, which was the reduction of suffering of the individual later in their life. The intent was only to produce a baby that would become a healthy adult.[53] There were now two individuals with the same disease on the two sides of the barrier—which results in severe similarity vagueness.

The threat to the barrier is clearer if we imagine the barrier that would need to be built lower on the slope to accommodate these applications if the somatic/germline barrier were to fall. The barrier would have to be between the individual who would eventually develop a disease, be they an adult or an embryo, and the species. However, biological reality makes a new barrier at that location impossible—the adults who eventually result from the modified embryos are going to reproduce, spreading the modifications to the species, unless all the resulting modified adults are forcibly sterilized, which would run afoul of other important values. If the somatic/germline barrier were to fall due to these arguments, it could not be replaced.

While proposed, these efforts to expand the acceptable acts of HGE below the germline barrier were not successful—a majority of the debate still assumed the barrier was in place. This was for a few reasons. First, there was no reason to risk any public or legislative controversy when the technology remained unavailable. At the time, scientists could not even get somatic HGE to work, so nobody would risk trying such an intervention to create a baby. Second, removing the barrier would open a huge range of acts below on the slope, which would have been seen as radical or revolutionary. More importantly, the barrier was still held in place by the values that these revolutionaries were ignoring. For the barrier to fall by making beneficence the only value, the defenders of the previous barrier would have to have become powerless, which was in process with the decline of the theologians but not yet complete. Therefore, despite developing the intellectual justification to take down the germline barrier, it remained standing, albeit with a road map for its destruction established in the debate.

The Emphasis on Autonomy Undermines the Germline Barrier

The most revolutionary challenge to the somatic/germline barrier in the debate begins in this era with the claim that the acts on the terrain on the downslope side of the germline barrier are not about the species but about human reproduction, because the only way to change the species is to gestate a baby. As a consequence, any limits on germline human reproduction violated the value of personal reproductive autonomy, which was an increasingly powerful value in this era. If broadly accepted, acts on essentially the entire slope would be allowed.

This was a more radical claim because autonomy would support acts across the entire slope of the debate from the germline barrier down, ignoring any distinction between disease and enhancement that the advocates of beneficence wanted to maintain. The terrain, which is defined by the target of the intervention, would change from species to baby—all the way down the slope.

The value of bodily autonomy was not really present in the debate during the original design of the barrier, and thus the barrier was not designed to resist it. For example, when the original barriers were created, doctors did not have to ask your consent for procedures and might not even tell you if you had a disease.⁵⁴ Abortion was illegal throughout most of the United States, with the 1973 *Roe v. Wade* decision still a few years away. *Roe* later established the idea that women should be able to decide autonomously whether to have an abortion early in pregnancy. In later years, pregnancy came to be thought of as reproduction, and all aspects of reproduction came under the logic of autonomy, at least for those in the public bioethical debate. As new technologies were developed that facilitated pregnancy, such as in vitro fertilization (IVF), as well as those that allowed women to avoid having children who would have certain genetic qualities, those too were justified by the value of autonomy. For example, women had the right to use chorionic villi sampling followed by abortion to avoid having a child with the genetic trait called Tay-Sachs disease.

Prominent voices in the debate began to consider the entire slope to concern reproduction and be governed by the value of autonomy. For example, in one of the most influential articles of the era,⁵⁵ John Robertson claimed that people have an fundamental autonomy right to germline HGE because "gene therapy on the embryo is closely tied to procreative choice.... The U.S. Constitution, it may cogently be argued, gives the parent the right to provide his or her children and their descendants with a healthy genome.... Properly understood ... the right to procreate includes a right to practice negative eugenics—to deselect harmful characteristics from future generations."⁵⁶

Another group of prominent authors wrote that "in pursuing their goal of helping individuals realize their own reproductive goals, medical genetics places more emphasis on respecting their patient's autonomy than almost any other medical specialty." Therefore, "it suggests that the primary question that clinical geneticists should ask ... is simply whether techniques for germ-line intervention will effectively improve their ability to respond to the reproductive health concerns and complaints of their patients."[57] This interpretation of autonomy would make the decision to use germline HGE as autonomous as any other reproductive decision.

Technology Changes the Terrain

The individual below the germline barrier identified by the advocates of the value of beneficence was purely hypothetical as the technology to modify a reproductive call or an embryo did not yet exist. In contrast, developments in reproductive genetic technologies resulted in actual germline babies of a sort, redefining the terrain of the slope and implicitly removing the germline barrier.

When the germline barrier was constructed, you could not influence the genes of your baby beyond the decision of who to mate with. But, technological improvements, combined with autonomy-based reproductive medicine, led to couples influencing the genetic qualities of their offspring through genetic *selection* technologies. Selecting the genetic qualities of a baby from among the possibilities also had the largely unacknowledged consequence of shaping, in however small a way, the species. For example, if you can select your child so that it is not a carrier of sickle cell anemia, when that child later reproduces the incidence of sickle cell in the population will be reduced.

These technologies were slow to develop. In the 1960s amniocentesis allowed a pregnant woman to determine whether the fetus she was carrying had Down syndrome. If so, she could select against the trait and elect to have an abortion. This was the first "genetic" test in reproduction, although it was restricted to gross chromosomal abnormalities.

In 1978 the first "test tube baby" was born by creating embryos in a dish for later implantation in the woman. This did not influence the genetic qualities of the offspring because the sperm and egg of the couple were joined, just like in standard procreation. There was no selection.

However, with those embryos in a dish, it was not long before it was figured out that the genetic qualities of those embryos could be accessed. Pre-implantation genetic diagnosis (PGD) was invented in 1989. It consisted of removing one cell from a number of eight-celled embryos in a dish. (This technique now diagnoses embryos at a later stage when there are more cells.) With this cell the genetic makeup of a baby that would eventually be born from that embryo can be determined. Embryos with desirable traits are implanted in the woman, and those with undesirable traits are discarded.

All those genetic diseases being focused on in the somatic gene therapy part of the slope (like sickle cell, beta thalassemia, ADA-deficiency) were also the diseases that couples were trying to avoid in their offspring using PGD. Therefore, people who knew they were carriers of a disease like sickle cell could use PGD to select a child that was virtually guaranteed not to manifest the disease.

As time moved on, more and more genetic traits could be identified and avoided in one's offspring using PGD. To the extent this technology was controversial, it was only so with those who were opposed to destroying embryos as well as the much smaller group of persons adhering to traditional Catholic natural law theory who would be opposed to destroying embryos *and* all of the precursor steps of removing gametes from male and female bodies.

Participants in public bioethical debate quickly realized that PGD produced a baby on the slope below the germline barrier. By intervening in what would have "naturally" occurred, a couple selected the genes of a baby that would be passed down to that baby's descendants and ultimately to the human species. PGD is then a type of germline HGE, albeit with the repertoire of genetic outcomes limited to single-gene traits understood by scientists and by the traits that those two parents could possibly produce on their own.

In 1991, bioethicist LeRoy Walters presciently predicted a scenario where a couple would like to use PGD but could not because they were opposed to destroying embryos or were incapable of producing embryos that did not have disease. If germline HGE were performed for this couple, the goal would be to produce an individual—a healthy baby—not influence the species: "in both of these scenarios, germ-line transmission would be a foreseeable but unintended side effect of a therapeutic procedure intended primarily to cure disease in an (embryonic) individual."[58] If the "primary" intent is to cure disease, then there is extreme similarity vagueness with curing disease above the barrier. Walters also exposed the design flaw in the germline barrier we have been discussing—PGD has the goal of influencing the genes of a baby (an individual), while *inadvertently* influencing the genes of all the descendants of that baby (the species).

Similarly, in the same year Zimmerman wrote that PGD was a strategy for germline intervention, saying "it is a matter of definition whether or not embryo screening and selection should be considered as 'germ-line therapy.'" He continued by saying it would be "the simplest, safest, and more reliable means to prevent a large faction of genetic disorders," and "the method of choice when a significant fraction of genetically normal embryos can be obtained following IVF."[59]

A Hypothetical New Barrier Downslope

Now that PGD was being widely practiced and was placing babies below the old germline barrier, a new barrier should have been built downslope if people wanted to avoid further slippage. Upslope of this new hypothetical strong barrier, in the approved zone of what was already happening in society, would have been what I would call "germline selection," and downslope below the barrier would be "germline modification." Above the barrier is "selection" because the influence on future generations comes from using PGD to select among embryos that the couple produces, which obviously limits the genetic effects to the gene variants that the parents could themselves produce. "Modification" means taking the reproductive cell of an adult and making a change regardless of what the parents themselves could have produced in a child. <u>I would argue that the distinction between modification and selection is not subject to similarity (or continuity) vagueness, so this would be a structurally strong barrier.</u>

However, a new barrier was not built—the fact that PGD jumped over the germline barrier was barely noticed. One reason is that PGD was a part of the abortion debate, not the HGE debate. Another was that germline *selection* was positively evaluated by most of the values that held up the old barrier, so it was not controversial. For example, germline *selection* through PGD did not really create social inequality because the parents could have possibly produced such a child without PGD. With IVF already demonstrated as safe, and the initial PGD uses having been conducted before the debate noticed, selection PGD also did not violate the value of non-maleficence. Additionally, selection was pretty consistent with the value of humility/hubris. One could not redesign the species with PGD of single-gene genetic diseases, and again the couple could have produced such a child by just having sex.

Perhaps counterintuitively, selection germline via PGD was also at best a mild affront to the values of following nature or God's will—as long as people consider IVF and selection "natural," which I recognize will not be agreed to by all. But, PGD would be "natural" or God's will because the couple could have produced such a child on their own through sex—the selection just changed the odds. Moreover, the degree of control over nature or God's will was limited in that you could only select from gene variants the couple themselves could produce. This is in contrast to modification, where any trait could be engineered into offspring.

Let us consider selection in light of some of the classic arguments holding up the germline barrier that value nature and/or God's will. Selection would be more consistent with philosopher Michael Sandel's concerns about not trying to overly control nature than would modification. I suspect Sandel would consider selection to be a mild form of control, but modification is a categorically different and more extreme type of control. Sandel sees the deepest danger of human-enhancement technologies "is that they represent . . . a Promethean aspiration to remake nature, including human nature, to serve our purposes and satisfy our desires. The problem is not the drift to mechanism but the drive to mastery. And what the drive to mastery misses and may even destroy is an appreciation of the gifted character of human powers and achievements." With children, he says, we must appreciate them as gifts and "accept them as they come, not as objects of our design or products of our will or instruments of our ambition. Parental love is not contingent on the talents and attributes a child happens to have." We must have an "openness to the unbidden."[60] I would argue that "design" is a better description of modification than selection.

A Christian, theological analog to Sandel's claim is that of Gilbert Meilaender, building on the work of Oliver O'Donovan, who makes a distinction between "making" and "begetting" children. In the Nicene Creed, the Son of the Father is "begotten, not made," which is meant to describe equality of being. He extends this to people being designed. "What we beget is like ourselves," he writes, "what we make is not; it is the product of our free decision, and its destiny is ours to determine." When a baby is made, "no longer then is the bearing and rearing of children thought of as a task we should take up or as a return we make for the gift of life; instead, it is a project we undertake if it promises to meet our needs and desires." "Making" results in children being considered more like objects.[61] "Making" is a better description of modification than it is of selection. Again, both Sandel and Meilaender would probably also see PGD as not being open to the unbidden or making instead of begetting.

But, one of the reasons that selection via PGD did not result in a new barrier is that it was not obviously or radically inconsistent with the values supporting the somatic/germline barrier.[62] Instead, the old germline barrier remained. There was little need to think in detail about these distinctions in the 1990s because any drive to conduct modification, and not just selection, abated. I see two reasons. One reason was, after many years of trying, scientists could still not get somatic modification HGE to work, suggesting that germline modification would always be impossible. Second, PGD selection allowed most couples who were carriers of genetic diseases to have genetically related children, obviating the motivation to conduct modification HGE. As a few bioethicists recognized at the time, "the prospect of pre-implantation screening—far less risky and more feasible than either of the gene therapy strategies already discussed—weakens the case for developing germ-line therapy by simply making it unwarranted for the majority of severe genetic disease to which it would first become applicable."[63]

So, while the foundations of the traditional somatic/germline barrier had been weakened, because we were allowing cases of selection below it, there was much less pressure to take it down given that the primary applications below the barrier that scientists actually wanted to do were much more easily done through selecting embryos. The best way to describe the situation using my metaphors is that the somatic/germline barrier slid down the slope to the terrain between selection and modification, and the acts of selection were acceptable, but nobody provided the value basis for the new location of the barrier. The final push on the somatic/germline barrier would wait for a few decades.

Weakening the Disease/Enhancement Barrier

The disease/enhancement barrier is primarily based on the definition of a disease, and whatever is not a disease is an enhancement.

But, what is a disease? This barrier worked well when everyone focused on the consensual cases of monogenic genetic disease far up-slope from the barrier. That is, everyone agreed in the 1950s and everyone agrees today that Tay-Sachs is a genetic *disease*. Everyone agreed in the 1950s—and everyone agrees today—that changing a baby to make them more intelligent is an *enhancement*.

However, our increasing knowledge of genetics over time revealed the design flaw in this barrier—it lacked a precise definition of disease—resulting in severe susceptibility to a similarity vagueness slippery slope process for applications closer to each side. As Eric Juengst wrote in 1997, "while the somatic/germ-line distinction is accused of lacking adequate ethical force . . . the conceptual line between these two classes of intervention is at least clear. The treatment/enhancement distinction, however, often seems in danger of evaporating entirely under its conceptual critiques."[64] This was a design flaw from which the barrier would not recover.

Even those who proposed taking down the germline barrier and relying upon the disease/enhancement barrier recognized that the latter was susceptible to slippery slope processes and probably would not hold. W. French Anderson, one of those who advocated ignoring the somatic/germline barrier to further beneficence, wrote that "even those of us who are the most enthusiastic proponents of gene therapy" have "a hesitancy about taking the first step" because of "the slippery slope." He continued: "Successful somatic cell gene therapy also opens the door for enhancement genetic engineering, i.e., the supplying of a specific characteristic that individuals might want for themselves (somatic cell engineering) or their children (germline engineering) which would not involve the treatment of a disease." His solution is to ground the barrier at "serious disease," but he quickly recognizes that this would fall prey to what I would call extreme similarity vagueness. Without defining "serious" or "disease" he pondered:

[W]hat distinguishes a serious disease from a "minor" disease from cultural "discomfort"? What is suffering? What is significant suffering? Does the absence of growth hormone that results in a growth limitation to two feet in height represent a genetic disease? What about a limitation to a height of four feet, to five feet? Where does one draw the line? Each observer might draw the lines between serious disease, minor disease, and genetic variation differently.[65]

He identifies many similarity vagueness mechanisms (e.g., serious vs. minor) and a continuity vagueness mechanism (e.g., four feet to five feet in height). In the end he defends the placement of his barrier by writing that "all can agree" on a condition that leads to "significant suffering and premature death." The later debate will demonstrate that his empirical claim about social consensus in defining disease is false.

Another set of authors who strongly endorse removing the germline barrier, and who argue that HGE should be ruled by the value of autonomy used in reproductive medicine, are also clear that the disease/enhancement barrier will be difficult to defend. They write that "if the client-centered ethos is strong enough to make medical geneticists take seriously parental assessments of the value or disvalue of their children's traits, the autonomy of clients to evaluate and change their own traits in preparation for childbearing—in short, to practice eugenics on a personal level—will be very hard to restrict."[66]

They gesture to the disease/enhancement barrier by stating that "medical geneticists' response has been that they, as health professionals, are only committed to facilitating reproductive choices prompted by the risk or presence of 'genetic disease.'" But, the authors acknowledge the "blurry edges of that concept."[67] The later debate will also show that health professionals lack agreement on the definition of disease.

The authors continue by pointing to how the increasing knowledge of genetics in their era was, in and of itself, making definitions difficult. The earliest diseases that geneticists talked about in the 1960s had a shared feature—you had or did not have the disease, and if you had the disease, you manifested symptoms almost immediately as a baby. But, the authors point out that genetic research has begun to identify "the genetic roots of an increasing range of predispositions and susceptibilities to disease" and diseases "that do not manifest themselves clinically until late in life."[68]

In 1986, Walters wrote that "within the past year, new markers for several major genetic diseases have been described, among them cystic fibrosis, polycystic kidney disease and Duchenne muscular dystrophy.... Applied to children or adults, the new tests will provide early notice of a tendency to develop a particular disease (such as atherosclerosis) or of the presence of a gene that will cause a late-onset disorder (such as Huntington's disease)."[69] It also became increasingly clear that these propensities toward disease were largely polygenic, eliminating the clarity of the early distinction between "diseases" as monogenic and enhancements as polygenic. This further reduced the differences between the two sides of the barrier.

Disease as Deviation from Normalcy

Once genetic discoveries made what constituted a disease nonconsensual, various participants in the debate attempted to define the disease/enhancement barrier so that it would not be susceptible to slippery slope processes. Following other developments in the philosophy of medicine, the most developed attempt to create a solid barrier was at "normalcy." The normalcy standard was well stated years later by the President's Council on Bioethics:

"Therapy," on this view as in common understanding, is the use of biotechnical power to treat individuals with known diseases, disabilities, or impairments, in an attempt to restore them to a normal state of health and fitness. "Enhancement," by contrast, is the directed use of biotechnical power to alter, by direct intervention, not disease processes but the "normal" workings of the human body and psyche, to augment or improve their native capacities and performances [70]

Similarly, the recent National Academies of Sciences, Engineering, and Medicine report on human gene editing defines enhancement as "changes that alter what is 'normal,' whether 'normal' for humans as a whole or 'normal' for that particular individual prior to being enhanced."[71] Philosopher Norman Daniels wrote that enhancements were "interventions that improve a condition that we view as a normal function or feature of members of our species."[72]

In the most basic version of this argument, normalcy is like a bell-shaped statistical distribution with "normal" being the bulk of the cases in the middle and "not normal" being the cases in the tails. It is then non-normal for an adult to be four or seven feet tall. It is normal for an adult to be five foot, six inches tall. To place this on the HGE slope, it is non-normal to have Tay-Sachs disease, and it is non-normal to have an intelligence like Einstein. If the disease/enhancement barrier were placed at normalcy, you could modify your child to make them normal on the Tay-Sachs trait (i.e., not have the disease) and could modify your child to give them normal intelligence if they were destined to be Einstein (not that anyone wants to reduce the intelligence of their offspring). You could not make your child like one of the cases in the tails of the distribution because that would make them "non-normal."

Early attempts by philosophers and bioethicists to define disease and enhancement as normalcy were implicitly based on solid design principles for avoiding slippery slopes—they tried to move away from subjective to objective definitions of disease. For those who defined the barrier this way, the terrain on the upslope side of the barrier was then acts which would return the body to "normal functioning," and on the downslope side, enhancement was going "beyond normal." This was objective because in principle normalcy was a statement about the entire population, independent of how an individual evaluated this information.

The Susceptibility of the Barrier Anchored at Normalcy

Attempts to shore up the barrier by defining disease as normalcy did not produce a barrier that was any less susceptible to slippery slope processes. The first problem was that an objective definition of normalcy is in tension with our individualist era where people have the right to define their own experiences. That is, "normalcy" assumes that one has to be within "species-typical range" in order to flourish. But, to put it bluntly, there are people who are not in the normal part of the bell curve—such as those with dwarfism—who do not want to be changed, do not think that they have a disease, and consider themselves to be flourishing quite well.[73] This raises the general problem that an objective measure does not necessarily match one's subjective sense of disease.

The second major problem for normalcy is that it is based upon the normative power of what currently *is*—on nature or God's design. Erik Parens notes that critics of enhancement tend to rely upon definitions that appeal to nature—to some variation on the idea that "whereas treatment restores normal or species-typical human functioning, enhancement does more than that." The values behind this type of reasoning are not promoted by the bioethics profession but are rather more likely to be found among the theologians who had been leaving this debate. These values had a short revival in the 2000s during the George W. Bush presidency when his bioethics commission championed the type of reasoning that held in bioethics before the bioethics profession took over. But, this was short-lived.[74] Without a large group promoting the value of following nature or God's will holding up the barrier, it is unlikely to stand.

A third major problem is that normalcy has become extremely susceptible to the continuity vagueness slippery slope process. Normalcy works better with categorical disease—a disease that you either have or do not have and which manifests itself similarly across all who have it. Tay-Sachs may be the prototypical example. But, many genetic traits exist along a continuous distribution. The most obvious is height, which forms a bell-shaped curve. Is an adult male who is five foot five normal? How about four foot ten? The same problem exists with a genetic propensity to heart disease. Is a 50% increased prediction of heart disease normal? 51%? How about a 3% increase? The same can be said about late-onset diseases with genetic influences. Alzheimer's at age 90? 80? Many "genetic diseases" now are described on a continuum, making them extremely susceptible to continuity vagueness slippery slope processes. Nobody would even try to put a propensity to Alzheimer's at age 50 above the disease/enhancement barrier and Alzheimer's at age 51 below the barrier.

A fourth problem was that the very first *somatic* therapy HGE clinical trials in the late 1980s hopelessly muddled the distinction between normality, disease, and enhancement in the debate. Given the debate in the 1970s, one would think that the first somatic gene therapy trial would have been to heal an inherited genetic disease like sickle cell.[75] However, it turns out that the first "human gene therapy" trial approved by the US federal government in 1989 was not for an inherited disease at all but for the genetic modification of normal immune cells in the body so that they could more effectively fight cancer.[76] To be overly dramatic, this created an enhanced, non-normal superhuman in the service of fighting a disease. (It is not normal to have these modified cancer fighting genes.) Participants in the debate in this era recognized this tension and identified applications on both sides of the barrier that would be made similar because they were both justified by "the proper goal of medicine." Juengst, pointing to these first somatic trials, wrote that the distinction between disease and enhancement

> dissolves in the case of using human gene transfer techniques to prevent disease when such interventions involve the enhancement of the body's health maintenance capacities. The argument is that to the extent that disease prevention is a proper goal of medicine, and the use of gene transfer techniques to strengthen or enhance human health maintenance capacities will help achieve that goal, then the treatment/enhancement distinction cannot confine or define the limits of the properly medical use of gene transfer techniques.[77]

Indeed, over the years, most of the somatic gene therapy research was not about genetic disease but about other medical goals, like fighting cancer.[78] Indeed, a 1997 editorial in the scientific journal *Gene Therapy* stated that "in spite of its independent origin and heritage, and whether anyone would wish to rename gene therapy, it is part of pharmacology."[79]

The original disease/enhancement barrier was not susceptible to slippery slope processes because there was no similarity vagueness between a single-gene rare disorder like Tay-Sachs and the polygenic trait called "intelligence." What was later found to lie closer to the barrier, like traits that are propensities to future diseases and traits that are not universally considered to be diseases (like deafness), destabilized the barrier. Is removing a 50% propensity to cancer fighting a disease or an enhancement? How about a 1% propensity? Or, making someone less susceptible to cancer than the rest of the population? Indeed, it was the first actual applications of HGE that began destabilizing the barrier, as these applications could be interpreted as creating non-normal enhanced humans... in the name of fighting disease. The intellectual grounds for knocking over this barrier were in place.

The Somatic/Germline and Disease/Enhancement Barriers at the End of the 20th Century

Barriers had been built in the early 1970s that allowed all participants to venture onto the slippery slope and agree to somatic gene therapy. For a while this was all quite clear and not subject to slippery slope forces. But, both barriers soon weakened. The somatic/germline barrier fell prey to the increasing dominance of the values of beneficence and autonomy as well as the design flaw that an individual baby exists on both sides of the barrier, resulting in a similarity vagueness slippery slope mechanism. What was needed was a justification for a selection/modification germline barrier, but such a justification was not developed.

The disease/enhancement barrier was largely undermined by increasing genetic knowledge that identified all sorts of "gray area" cases of disease—a gray area that allowed for claims that an application upslope of the barrier was "just like" an application downslope of the barrier. Scholars have spent a good amount of time trying to make such a distinction hold, such as accounting for both biological normalcy and cultural definitions of disease at the same time, but these end up producing barriers no stronger than that which we have already considered.[80] Other scientific innovations led to the first somatic gene therapy trials not being for inherited genetic disease at all, but for enhancing the cells in people's bodies to fight diseases like cancer, further muddling the justification for the barrier.

The barriers still stood in the debate as hollowed-out remnants of their previous selves but without replacements. Both barriers were largely kept standing due to the support of the value of non-maleficence. That is, scientists would not go beyond these barriers because to do so was not yet demonstrated to be safe, as nobody knew how to make germline changes or how to create the sort of enhancements that people were interested in. Soon, a major technological innovation would suggest that these technologies would be safe, providing the impetus to knock over both weakened barriers.

3
The CRISPR Era, the National Academies Report, and the Median Trait Barrier

The historical narrative in the previous chapter stops at the end of the 20th century, and the debate did not significantly advance in the first decade of the 21st century, as there was very little motivation to push against the existing barriers. One reason was the inability to get the long-approved-of somatic gene therapy to work for genetic diseases in people's bodies. In 1999 a volunteer in a trial died from a reaction to the viral delivery device, not to gene modification itself. In 2000 three patients were cured of an immunodeficiency disorder, but the therapy caused a leukemia-like disease in two of the 11 patients. A 2003 review of progress toward somatic gene therapy noted that early enthusiasm for the technology "rapidly waned as clinical trial after clinical trial failed to show efficacy" because the early viral delivery vehicles for the genes failed to deliver the genes properly. The authors stated that "substantial challenges . . . remain before gene therapy can truly fulfil all of its promises."[1]

Over 10 years later, the story remained the same. In 2015, another scientist, writing in *Nature*, while now hopeful about somatic gene therapy, bemoaned the lack of progress to date. He stated that "gene therapy has long fascinated scientists, clinicians and the general public because of its potential to treat a disease . . . by replacing a malfunctioning gene within the cells. . . . As simple as the concept sounds, the hurdles to put it into practice are daunting." He continued, "several gene-therapy trials have been performed in the past two decades for inherited diseases, cancer and chronic infections, but only a few reported clear clinical benefits and in some, individuals experienced severe adverse events related to the vectors" that deliver the genes.[2]

The inability to get somatic gene therapy to work for extreme diseases (which allowed for an acceptable risk-benefit ratio) meant that there was no one pushing toward the more ambiguous diseases near the disease/enhancement barrier. One reason why "first-in-human" experiments are limited to those with serious diseases is that the benefit a patient would receive is strong, which outweighs the high risk of early trials. Extreme diseases, like severe combined immunodeficiency (SCID) and beta thalassemia, remained the targets of scientists' genetic disease efforts.[3]

There was also no pressure from the scientific community to make a genetic change that would express in children and spread to all of that child's progeny (i.e., germline modification HGE) when they could not get somatic gene modification to work. If it could not be made to work on an existing adult, nobody was going to try creating a not-yet-existing human in order to satisfy the desire of parents to have a genetically related child.

The motivation to test the strength of the barriers was also lacking due to the further development of PGD. *If* germline modification were ever to be attempted, it would first start with extreme diseases that could justify the risk of an early trial, and with a relatively simpler single-gene disorder. For example, there are couples who are carriers for diseases like sickle-cell anemia who want genetically related children, but who are at risk of having a child who expresses the disease. But, there was no reason to engage in modification when they can achieve the same goal by using PGD (selection). Moreover, PGD had already pushed into the terrain near the disease/enhancement barrier with the ability to identify embryos susceptible to adult-onset disorders and traits like deafness.[4] The further development of traits for which PGD could be used strongly limited the potential patient base for therapeutic germline modification HGE. Finally, the germline barrier was still the legal standard throughout Europe where most countries had banned germline modification via laws or treaties.[5]

The Barriers at the Beginning of the CRISPR Era

In June 2012 the journal *Science* published an article showing that genes could be "edited" using the CRISPR-Cas9 protein.[6] Exactly how this technology works does not matter to this book, but many excellent summaries are available.[7] What is important, is that what had always stopped genetic modification from working was the inability to perform site-specific modifications. For example, the viral vectors that delivered the new genes inaccurately placed the new genes on the genome, causing a host of problems, such as cancer. With the CRISPR technology, a gene could be cut out of the genome at an exact desired location and replaced. Metaphorically, the previous methods were like dropping a piece of paper with a word on it on a manuscript page and seeing where it would land. The new method was called "gene editing" because if a word was misspelled, you could search and replace as you would with a word processing program. For the first time, scientists had the ability to accurately engage in genetic modification (now commonly called gene editing).

Two scientists influential in the development of CRISPR, Jennifer Doudna and Emmanuelle Charpentier, stated that "CRISPR-Cas9 has triggered a revolution in which laboratories around the world are using the technology for innovative applications in biology." The last line in their editorial stated that the technology "will also be critical for applications of the technology in human gene therapy."[8] After years where it appeared that HGE would always be impossible, it was now seemingly immanent.

Regathering at the Germline Barrier

Surprisingly, the first attempts to use CRISPR for humans was for germline not somatic modification. The news broke in April 2015 that Chinese scientists had edited the genomes of human embryos, which would be the first step in germline HGE, but had not attempted to have a woman bring them to term. (The lead author claimed that the article reporting the research had been rejected by both *Nature* and *Science* on ethical grounds.) It was reported that at least four research groups in China were working on gene editing human embryos.[9]

Earlier rumors of the Chinese efforts had restarted the HGE debate, and participants returned to their previous positions at the barriers on the slope.[10] There was a flurry of activity in public bioethical debate, most of which centered on the germline barrier, still in place after all of those years. Since most of what was holding up the germline barrier was that it was considered inaccurate and unsafe downslope (non-maleficence), making gene editing accurate and safe threatened to knock over the barrier. After all this time, would anyone remember the other values that held up the barrier?

Many scientific associations and groups of scientists released position papers on HGE, mostly defending the germline barrier using the value of non-maleficence. These were largely groups focused on somatic therapy, so they had no interest in germline. An example was the director of the National Institutes of Health (NIH) who released a statement stating that the "NIH will not fund any use of gene-editing technologies in human embryos. The concept of altering the human germline in embryos for clinical purposes has been debated over many years from many different perspectives, and has been viewed almost universally as a line that should not be crossed." The NIH also said that crossing the germline would result in "serious and unquantifiable safety issues, ethical issues presented by altering the germline in a way that affects the next generation without their consent, and a current lack of compelling medical applications."[11]

The National Academies of Science, Engineering, and Medicine released a report during this time on a somewhat different topic—mitochondrial transfer in embryos, which could result in genetically modified embryos. The report quite conspicuously kept upslope of the germline barrier by allowing mitochondrial transfer for only male embryos to avoid transmission of the genetic change to subsequent generations.[12]

In August 2015 the American Society for Gene and Cell Therapy and the Japan Society of Gene Therapy released a statement saying that the new technology would contribute to somatic gene therapy, then defended the germline barrier, justified by the then-dominant value of non-maleficence. They consider the "safety and ethical concerns" about germline HGE

> to be sufficiently serious to support a strong stance against gene editing in, or gene modification of, human cells to generate viable human zygotes with heritable germ-line modifications. Even with technical advance that may eventually solve the safety and mosaicism problems, our Societies conclude that there are not ethically acceptable ways to conduct embryonic gene editing or other germ-line modifications because the results of such experiments are not susceptible to long-term evaluation in a scientifically reasonable time scale. For these reasons, our societies support a strong ban on human germ-line gene editing or other germ-line genetic modifications unless and until these technical and ethical problems can be solved, broadly and deeply discussed, and societal consensus reached.[13]

Similarly, a group of somatic gene therapy researchers associated with the Alliance for Regenerative Medicine, which represents life sciences companies, assumed that "studies involving the use of genome-editing tools to modify the DNA of human embryos will be published shortly." They were fearful of "a public outcry" that "could hinder a promising area of therapeutic development, namely making genetic changes that cannot be inherited." They pointed to the slippery slope, writing that "many oppose germline modification on the grounds that permitting even unambiguously therapeutic interventions could start us down a path towards non-therapeutic genetic enhancement. We share these concerns." They wrote that "patient safety is paramount among the arguments against modifying the human germline" because it "could have unpredictable effects on future generations," and call for a "voluntary moratorium" on germline applications.[14]

Non-scientists used a few more values to defend the barrier. The politically left Center for Genetics and Society released an open letter calling for "a prohibition on reproductive human germline modification."[15] They stated that "experiments with human germline intervention could lead to miscarriage, maternal injury, and stillbirth. . . . Other harmful consequences of germline modification might only present themselves in subsequent generations." However, unlike the scientific groups, in addition to non-maleficence they also invoked the value of equality or fairness, with the acts above the germline barrier supporting and the acts below working in opposition to that value. They point to the political liberal's *Gattaca* dystopia at the bottom of the slope when they wrote that "permitting germline intervention for any intended purpose would open the door to an era of high-tech consumer eugenics in which affluent parents seek to choose socially preferred qualities for their children. At a time when economic inequality is surging worldwide, heritable genetic modification could inscribe new forms of inequality and discrimination onto the human genome."[16]

In contrast to the United States, continental Europe has long had additional values that justify the barrier, largely due to the eugenics movements of the Nazi era which exposed the dangers of viewing people's value as derived from their genetic qualities. So, for example, germline *modification* is banned through much of Europe on the grounds that it would violate human dignity. Therefore, when CRISPR emerged, the germline barrier in Europe was still supported by values other than non-maleficence.

For example, the International Bioethics Committee of UNESCO published a report in 2015 which argues that germline modification would "jeopardize the inherent and therefore equal dignity of all human beings and renew eugenics." This is the group where the member states adopted the Universal Declaration on Bioethics and Human Rights in 2005, which states that "the human genome is part of the heritage of humanity," and therefore outlines "rules that need to be observed to respect human dignity, human rights and fundamental freedoms."[17] Concerns about dignity harken back to the Brave New World version of the bottom of the slope where humans are considered to be more object-like.

Organizations representing the German scientific establishment also released a statement. While they did not state a substantive, non-safety reason to defend the germline, they did gesture to the idea that even if it were proven safe, it might not be right to do it: "Even if the efficiency, specificity and safety of genome editing do one day meet the requirements for the responsible application of the techniques and can be used to stop severe hereditary diseases being passed on to subsequent generations, it must be clear whether and under what circumstances intervention in the germline is acceptable. A moratorium should ensure that—in the future, too—these methods are dealt with safely, transparently and in accordance with ethical principles."[18]

US Scientific Establishment Questions the Need for the Germline Barrier

It is striking how weak the somatic/germline barrier had become in the United States. When created it was supported by the value of following nature or God's will, which positively evaluated the acts above the barrier and negatively evaluated the acts below the barrier. The value of social equality had also supported the barrier, with equality above and inequality below.

Now, at least for the scientists, pretty much the only value left holding up the barrier was non-maleficence.[19] This value cannot support a strong barrier on its own because our developing scientific knowledge continuously redefines the acts just below the barrier from "not safe" to "safe," allowing "safe" to be on both sides of the barrier. That is, "safe" is subject to extreme similarity and continuity vagueness, requiring continuously moving the barrier down the slope to the location where our knowledge ends.

While the somatic gene therapy scientists in the United States were calling for the maintenance of the germline barrier—albeit justified only by non-maleficence—a number of elites in the US scientific establishment, working with bioethicists, were suggesting a re-examination of whether the germline barrier was needed. Around the time of the Chinese embryo experiments, a group of some of the most influential American genetic scientists and bioethicists met in Napa Valley, California, to discuss germline modification. The group included, among others, David Baltimore and Paul Berg (who had been central in the Asilomar debates of the 1970s),[20] Jennifer Doudna (one of the developers of the CRISPR process), George Church (a well-known scientist in genetic research), and Henry Greely and R. Alta Charo (prominent bioethicists). Their statement was published in April 2015 in *Science* magazine.

If they had strongly believed in the germline barrier, they would have simply reasserted the traditional arguments for maintaining it. Instead, they proposed a "prudent path forward for genomic engineering and germline gene modification."[21] They too focused on non-maleficence but simply concluded, unlike the other groups, that acts downslope of the somatic/germline barrier would soon be safe, making what they saw as the primary justification for the germline barrier moot.

This group had the attention of international institutional science and contributed to organizing a three-day international summit under the auspices of the National Academies of Science, Engineering, and Medicine; the Chinese Academy of Science; and the UK's Royal Society. The summit consisted of presentations, not deliberations. At the end of the summit, the organizing committee, made up of 10 scientists and two bioethicists, released a statement calling for the continuation of "basic and preclinical" research on editing human sperm, eggs, and embryos, as long as those entities were destroyed and not implanted in a woman. They reiterated the social consensus that somatic gene editing for disease is morally acceptable and should be treated like any other experimental medical treatment.

They also identified six issues of concern with germline modifications of humans. The first four were variations on risk or harms that our present lack of knowledge could cause, but, the authors of the statement implied, with advancing knowledge these concerns could soon be overcome. The fifth concern was that enhancements to subsets of the population could "exacerbate social inequalities or be used coercively," which is the one value supporting the barrier beyond non-maleficence traditionally used by contemporary liberals in the debate. The sixth was a vague catch-all phrase summarizing all ethical questions at hand—"the moral and ethical considerations in purposefully altering human evolution."[22]

The authors go on to say that the clinical use of germline gene editing should not be done "unless and until" the "safety and efficacy issues have been resolved," and there is "broad societal consensus about the appropriateness of the proposed application." Again, in contrast to the previous consensus against any germline modification, this presumes that some proposed applications will be acceptable and others not, and that therefore the germline barrier itself is not morally significant. Moreover, they say that "as scientific knowledge advances and societal views evolve, the clinical use of germline editing should be revisited on a regular basis." That is, if there is to be a barrier at all, it will be justified by the value of non-maleficence and will therefore be a sliding barrier as our knowledge increases and as slippery slope processes lead societal views to "evolve."

They continued by saying that the "international community should strive to establish norms concerning acceptable uses of human germline editing . . . in order to discourage unacceptable activities while advancing human health and welfare." In other words, human germline gene editing should go forward, but should distinguish between the acceptable forms, which promote health, and the unacceptable forms, which, we are invited to infer, promote enhancement. This is a vague gesture to the disease/enhancement barrier, still standing in its weakened form.

It might be surprising that this group publicly advocated pulling down the germline barrier with no in-depth examination or deliberation. However, if the only value supporting the barrier is non-maleficence, and scientists are saying that germline is about to become safe, it made no sense to defend the barrier. Safe applications of HGE would be on both sides of the barrier, resulting in similarity and continuity vagueness, making the defense of the barrier impossible. Implicitly, the only question was whether there were any barriers further down the slope.

Earlier it had been announced that the groups that had sponsored the summit would engage in an in-depth study of the ethics of HGE in the CRISPR era, despite what could be interpreted as a premature conclusion by the summit organizers. A committee was formed, which met five times between December 2015 and September 2016 and then released a report. It is important for transparency to report that I was a member of the committee.[23]

I will closely examine their report for what it says about the past and future barriers on the slippery slope. This close examination is warranted for a few reasons. First, <u>in the American scientific and bioethical context, reports from governmental and</u>

pseudo-governmental commissions—especially those seemingly endorsed by institutional science—have a strong impact on policy. For example, the first U.S. government commission report on HGE, the 1983 report *Splicing Life*, had a direct impact on policy at NIH and on which HGE studies were allowed.[24] The National Academies of Science, Engineering, and Medicine (NASEM) is one of these pseudo-governmental entities,[25] and the reports are expected to have impact on policy—particularly on the executive branch agencies controlled by scientists.

Second, reports like these are also considered to be summary statements that consolidate the public bioethical debate to that point. In the earlier HGE debate, *Splicing Life* had a powerful impact on the debate from that point forward.[26] The 2017 NASEM report also appears headed toward being a touchstone for the HGE debate going forward.

Thirdly, the conclusions of the NASEM report and others like it are given great intellectual authority. For example, as previously mentioned, the Chinese scientist who recently facilitated the creation of the germline enhanced twins justified his actions to his hospital's ethics committee by citing the NASEM report and claiming that the NASEM had approved editing human embryos for serious disease.[27] (A close reading of the NASEM report would not support what he had done.) In turn, the organizers of the Second International Summit on Human Genome Editing where he announced the experiment cited the NASEM report to condemn his actions.[28]

The NASEM Report Advocates Replacing the Somatic/Germline Barrier

The NASEM report repeats the long-standing consensus that somatic gene therapy is ethically meritorious, and it primarily discusses how we would know whether it is safe to proceed. More importantly, reflecting the dominant views of bioethicists in the HGE debate, the NASEM advocated pulling down the somatic/germline barrier and allowing couples to modify the genes of their child if 10 conditions are in place. Two of the conditions are that a public debate must occur and that oversight mechanisms must be in place to limit applications to those the NASEM endorses.[29] Three concern safety (non-maleficence).[30] Another is "absence of reasonable alternatives" for the couple wanting to avoid disease, which is not really a condition without a definition of "reasonable," which is not provided. The final three conditions are relevant for the barriers I have been discussing. Germline would be restricted to "preventing a serious disease or condition"; "editing genes that have been convincingly demonstrated to cause or to strongly predispose to the disease or condition"; and "converting such genes to versions that are prevalent in the population and are known to be associated with ordinary health with little or no evidence of adverse effects."[31] I will call this the NASEM barrier, which is a version of a disease/enhancement barrier, of which more below. But for now the interesting question is how the report advocated removing the somatic/germline barrier.

The Slippery Slope Across the Barrier

The germline barrier was taken down through a similarity vagueness slippery slope mechanism. As in all arguments against barriers by influential authors I described in the last chapter, the slippery slope across the barrier begins with extending the value that justifies upslope applications to the territory downslope of the barrier. In this case, it is quite striking, given the history of the debate, that the value of beneficence is not used at all in the argument for taking down the germline barrier. But wait, you are probably thinking, the word "disease" appears in two of the conditions and "ordinary health" in the third: How can I say this is not about beneficence? I will go slowly.

First, the report makes clear that you cannot produce beneficence by substantially reducing the number of "bad genes" in the total human gene pool through gene editing, unless technologically mediated reproduction becomes widespread, and the old-fashioned and more fun way to make babies becomes much less common. In the more professional cadence of the group, the report states that "germline genome editing is unlikely to be used often enough in the foreseeable future to have a significant effect on the prevalence of these diseases."[32] That is, until we get to Gattaca at the bottom of the slope, where most reproduction is guided with knowledge of the parents' genetics, germline modification is only going to have a direct impact on the descendants of the modified, which would remain a tiny portion of the population of the world.

More central to the report is that germline modification is not justified by beneficence toward the children that are born. Indeed, the "heritable genome editing" chapter begins with the following sentence, which is not about beneficence per se: "For prospective parents known to be at risk of passing on a serious genetic disease to their children, heritable genome editing may offer a potential means of having genetically related children who are not affected by that disease—a desire shared by many such parents."[33]

This statement accurately reflects <u>increasing acknowledgment in the debate that a germline modification to stop an embryo from having a "disease" is not actually about disease, health, or the reduction of suffering, but about parental desire for genetically related children</u>. The arguments of the 1990s implied that the only way to avoid children getting genetic diseases was to genetically modify them or their parents, but two people who are carriers for a genetic disease can have a child in many other ways that would avoid disease. The report repeatedly describes these alternatives.[34]

If a couple knew they were carriers of a genetic disease like sickle-cell (knowledge that is a prerequisite to using germline modification or selection through PGD), they could remain childless or adopt a baby or child. Those were the options available in the 1960s and remain available today. But, today they could also use a sperm donor or use someone else's surplus embryo left over from an IVF attempt and have a child without disease. If they were going to go through the physical risk and difficulty of extracting eggs for IVF in order to have the genes of their embryos edited, to say nothing of the great unknowns with embryo modification, it would be so much easier to start a family by not extracting eggs and instead using a donated embryo left over from someone else's IVF procedure. This would provide a healthy baby and also allow women to have the experience of pregnancy, compared to adopting an existing baby. Therefore, it is important to be clear that germline gene editing as endorsed by the NASEM is not about relieving disease per se, but it is about satisfying the desire of couples who are carriers of genetic diseases to have a genetically related child.[35]

So, beneficence was not the value that moved over the barrier. Instead, the new dominant value of autonomy alone justified removing the germline barrier. They wrote that "the desire to have genetically related children may arise from a variety of factors," and "precluding access to this technology could be regarded as limiting parental autonomy." "Indeed, some people feel they have a religious or historical mandate to have genetically related children." That is, if people want to have children with a particular trait—genetic relatedness—then this is their right to do so, absent a case for how their actions would harm others.[36] In actuality, the committee did come up with limits on germline reproductive autonomy for enhancements, as I will discuss below.

The First Similarity Vagueness Mechanism

It is informative for the future of the debate to see precisely how the germline barrier was felled, which was by particularly strong similarity vagueness mechanisms. The argument was that the applications on either side of the somatic/germline barrier were indistinguishable, and since we approve of the upslope application, we must approve of the downslope application. Recall that the barrier was located on the slope between selection (PGD), which was widely practiced, and modification (editing), which was not acceptable. This position on the slope was never justified by connecting it to values.

The upslope application—currently above the barrier and thus ethically meritorious—is for a couple who are carriers for a disease such as sickle-cell, who want a genetically related child who is healthy, so they engage in PGD and *select* an embryo that will not manifest the disease. Below the barrier are a couple with the same goal but who cannot use PGD, so they would need germline modification with gene editing.

The report extensively discusses the couples on either side of the barrier—those who can and cannot use PGD for the exact same reproductive goal—and it is argued that these two groups are so similar that no distinction (no barrier) can be made between them. Specifically, there are couples with particular genetic diseases for whom, if they tried PGD, "all or a majority of embryos will be affected, rendering PGD difficult or impossible." An example would be dominant late-onset genetic diseases, such as Huntington's disease, where one parent will be homozygous for the mutation. In that scenario, "all embryos would carry the dominant disease-causing allele that would cause the disease in the children, so PGD is not useful." This is one group, and another is those where "only one in four embryos would be free of a disease-causing mutation. Those unaffected embryos could be identified by PGD, but the number of embryos potentially available for implantation would be significantly reduced."[37]

The report also extensively discusses the possibility that while people who are carriers of a disease can use PGD, those who are homozygous (typically meaning they "have" the disease) cannot use

PGD, because all the embryos they produce would "have" the disease.[38] Finally, the report identifies additional women, also trying to have genetically related children free of disease, who cannot use PGD and are thus unjustifiably below the germline barrier. These include those who have mutations that compromise fertility, those with external factors like cancer treatments that reduce ovarian reserves and thus limit the odds of finding a healthy embryo via PGD. This section concludes: "In all of these situations, if it were safe and efficient to use heritable genome editing (e.g., in gamete progenitors) to correct the mutation, this alternative might be preferred by prospective parents who otherwise would be considering PGD."[39] This sentence perfectly describes the slide down the slope.

For the germline barrier at selection vs. modification to survive, someone would have had to come up with values that support it those many years previously when PGD was invented. Now it had to simultaneously distinguish between the two types of women who want germline modification—between the women who can use selection PGD and those who cannot use PGD so they want modification. How to distinguish these women? Those with the disease vs. carrier of the disease? With a body that produces a lot of eggs vs. with a body that does not produce a lot of eggs? Someone who fell in love with someone heterozygous for the disease vs. someone who fell in love with someone homozygous for the disease? In contemporary culture, and certainly in public bioethical debate, there are no values in use that can make these distinctions. A barrier at this location, between selection and modification, cannot be supported. In Chapter 1 I noted that a barrier was less likely to fall when the argument against it is radical, claiming a large amount of terrain below. In contrast, a barrier is more likely to be taken down if the value just barely slips over the top, covering a very limited number of applications just below the barrier. That was the case here. The report ends the section discussing those who cannot use PGD by stating "the number of people in situations like those outlined above might be small, but the concerns of people facing these difficult choices are real."⁴⁰ The report never states how many of these couples there are, but a presentation at the NASEM summit meeting suggests that the number of couples in the United States who could not possibly use PGD at all is in the hundreds.⁴¹

It was therefore argued that the practice of germline modification would not become widespread—the values that opposed it were not openly rejected—but instead an exception was to be carved out for only the few hundred couples who cannot use PGD. This small move across the barrier would not be controversial to anyone who supported PGD.

The Second Similarity Vagueness Mechanism

The report contained another similarity vagueness argument that puts equivalent acts on both sides of the germline barrier. This mechanism receives much less emphasis but is important to examine. At various points in the report, taking down the germline barrier is supported because PGD requires "discarding affected embryos, which some find unacceptable."⁴²

Therefore, on the upslope of the germline barrier, now located between germline selection and modification, is the couple who are carriers for a genetic disease like sickle-cell who use PGD because they are not opposed to destroying embryos. On the downslope of the barrier is the couple who are carriers for the same genetic disease but who cannot use PGD because they are opposed to destroying embryos, so they need to use modification. These two applications on the two sides of the barrier can only be separated by the couple's views of embryonic life, which, like the distinction between women who can and cannot produce enough embryos for PGD, is not a barrier that can be justified by any values in this debate. That is, on what grounds would we treat these two couples differently? What is interesting or ironic about this argument is that breaching the germline barrier to accommodate pro-life parents may well appeal to the same "bioconservative" constituency that has supported the germline barrier in the past.

Indeed, removing the germline barrier would result in the only form of reproductive intervention that would be consistent with traditional Roman Catholic teaching (called the magisterium). The official Catholic position is that anything that "divorces procreation from the conjugal act" is against natural law.[43] Therefore, contraception, abortion, removing embryos from the body for IVF, removing sperm from the body, destroying embryos—they are all illicit. (Of this list, only concern with abortion and destroying embryos is found in conservative Protestant theology.) So, while conservative Protestants may find repairing an embryo to be acceptable or even desirable, for traditionalist Catholics, getting that embryo out of the body in the first place would still be a violation of the magisterium. However, one approach to germline editing is to alter the spermatogonial stem cells that make sperm, and then put the stem cells back in the man. The stem cells would then generate mature sperm, and a couple could engage in the conjugal act and produce a genetically modified baby. According to one Catholic theologian, this would "bring the process more in line with magisterial teaching on sexuality."[44]

The committee clearly does not share this concern with embryonic life, given that there is an entire chapter justifying research on and the destruction of embryos solely for the purpose of developing our scientific knowledge and the efficiency of reproductive technologies. Rather, the concern with embryonic life is justified by the value of autonomy—there are couples for whom this is important, and their views should be respected.

In sum, the bioethics community had been pushing on the germline barrier since the 1980s, and finally an established, influential commission has advocated knocking it over. To recap our history, "influencing the genes of future generations" was the original location of the barrier. PGD later changed the location of the barrier to "influencing the genes of future generations through *selecting* the genetic qualities of children" on the upslope, and "influencing the genes of future generations through *modifying genes*" on the downslope. However, this shift in the location of the barrier was barely noticed and was certainly not publicly justified by connecting it to new values.

The germline barrier, in its new location, was knocked over primarily because autonomy became the primary—or even only—value considered. If the value is autonomy, there is extreme similarity vagueness between essentially identical sets of prospective parents on both sides: those who were carriers of genetic disease who could use PGD and those who could not, and those who thought PGD is morally acceptable and those who did not. However, the NASEM did not want to open the slope all the way down to Gattaca or the Brave New World, but rather advocated for another barrier lower on the slope.

The NASEM's Proposed Barrier

The NASEM barrier is located on the slope at "serious disease." To repeat the first two conditions on germline modification, germline applications will be restricted to "preventing a serious disease or condition" and to "editing genes that have been convincingly demonstrated to cause or to strongly predispose to the disease or condition." These refer to the previous disease/enhancement barrier, which the committee has been criticized for depending upon.[45] My interpretation is that they thought the traits that will be targeted for many years would be consensual diseases like sickle-cell, so they relied upon the disease terminology for short-term clarity, even if it cannot hold in the long run.

Indeed, the report focused on the instability of the term "disease," making it clear that the old disease/enhancement barrier cannot hold. They write that "many discussions of the ethics of enhancement have been based on contrasting the concepts of 'therapy' and 'enhancement.' However, given the evolution of the role of the physician over the past several decades from a healer of the sick to a promoter of health through preventive measures, the therapy-enhancement duality needs to be modified to accommodate a wide range of preventive interventions, such as vaccines, that are neither therapy nor enhancement but blend into each at the edges." They also acknowledge that the definition of disease is a social construct that shifts with society and with time. For example, "everyone would agree that the manifestation of Tay-Sachs disease is not normal and constitutes a disease, but opinions differ as to whether genetically caused deafness should be considered a disease." Moreover, homosexuality was, until recently, considered to be a disease.

Repeating the conclusion in the debate up to that point, they also noted that "the discovery of variants that simply increase the odds of developing a disease and others that are associated with diseases whose onset is in later life also has blurred the previously bright line demarcating 'disease.'" To round out their review of why the disease/enhancement barrier was not stable, they write that "the greatest challenge for the normality standard came from some researchers considering what might best be called enhancements for the purpose of relieving disease. This enhancement would not correct errors, but rather instill traits that some lucky minority of humans already have, such as by enhancing immune function or adding cellular receptors to capture cholesterol. . . . Such alterations . . . complicate the distinction between therapy and enhancement."[46] Indeed, an enhancement to combat disease was exactly what was supposedly done with the twins born in the 2018 Chinese experiment.

So the NASEM barrier will not hold beyond the first clinical trials of the most agreed upon diseases like sickle-cell. We are then left without a functioning barrier on the slope. For those who want to avoid the slide to the bottom, a long-term replacement needs to be found, and the NASEM report provides the materials with which to build that barrier.

The Prevalent Variant Barrier

The "prevalent variant" barrier can be justified further downslope using what is implicit within what the NASEM committee writes. A new barrier needs values to support it. As one would expect from the contemporary debate dominated by the arguments of the bioethics profession, while the committee wrote about many values, they only *used* the values of autonomy, beneficence, non-maleficence, and justice in their conclusions. More so than most bioethics commission reports, the NASEM report emphasized the value of justice or fairness, sharing the social liberal's conception of the slope where Gattaca is at the bottom—a world where people are socially defined by their genetic traits and all social interactions are structured by the genetic category the person is born into, resulting in a society of "genetic haves" and "genetic have nots."

This conception of justice was articulated in the report's discussion of enhancement, which followed the group's inability to define disease.[47] "Enhancement" is therefore discussed separately from disease and is defined in one place as "a change from—indeed an improvement upon—an existing condition."[48]

Despite using a version of the disease/enhancement barrier for their explicit barrier, they clearly suggest that they are laying the groundwork for a barrier below it for a later stage of the debate. They write that "while both the somatic/germline and disease/enhancement distinctions have been useful, they (like most categories) are imperfect. Some commentators have focused instead on the effect of an intervention and whether that effect is 'fair.' "[49] Fairness (justice) is the value that the prevalent variant barrier can be built upon.

Fairness has two requirements. One is that "germline modification could be used to create a level playing field for those whose traits now put their children and descendants at a disadvantage."[50] That is, those whose children would be disadvantaged due to disease or even some non-disease trait like intelligence, could be given an "equal opportunity" by bringing them up to what others have called a "genetic decent minimum."[51] The second requirement of fairness is to *not* allow people to use genetics to obtain social advantage. They write that "a problematic enhancement is one that confers a social advantage beyond that which an individual possesses by fate or through personal effort, and that does not benefit the rest of society in any way or undermines the implicit goals of a competition. Using equality of opportunity and societally useful inequality as guides may help distinguish those forms of enhancement that might generally be tolerated (assuming the risks are proportional to the benefits) from those that would be more controversial."[52]

In their report, the value of fairness is not defined in a way that could be used to build a barrier resistant to slippery slope processes. However, a strong barrier can be built—supported by the value of justice or fairness—by using the final, formal condition that the NASEM puts on HGE, which is that HGE should be restricted to "converting such genes to versions that are *prevalent* in the population and are known to be associated with *ordinary* health."[53] "Ordinary" signals the intent to allow for upgrades to a level playing field but not give advantage. That is, it is normal to not have a disease, and you cannot gain advantage by becoming ordinary. Critically, "prevalent" is the operationalization of this idea in a way that can be used to define a barrier—a prevalent variant will produce an "ordinary" human body at exactly that trait. "Trait" means a feature of the organism—in this case the human body.[54]

The concepts of prevalent variant, ordinary trait, and disease are all linked. First, at least for the monogenic diseases the committee focuses upon, a prevalent variant is by definition going to produce an ordinary trait in the body, not disease. Indeed, if a prevalent variant were bad for the body, it would have been weeded from the human gene pool via evolution, and scientists assume that being prevalent in the population indicates a variant is not associated with a disease. For example, genetic counselors consider a frequency of over 5% in the population "stand alone support" for a variant being benign for a rare Mendelian disorder.[55] Moreover, basic cultural sociology would conclude that for something to be considered a "disease" not everyone can have it—it cannot be prevalent. If everyone had a trait, even if it resulted in bodily decline (like aging), we would consider it part of normal human function.

A prevalent variant produces an ordinary trait, so using a prevalent variant simultaneously allows raising people to ordinary (e.g., to not have a disease), as well as not allowing a trait in the resulting human that would give them advantage. Therefore, putting a barrier at "prevalent variant" is supported by the value of fairness or justice because it allows, upslope, the upgrading of the traits of disadvantaged individuals to ordinary. This cannot be used to give traits that provide social advantage because you cannot gain advantage by becoming ordinary. Downslope, in the not allowed terrain, would be providing a non-prevalent variant that could generate a trait that would give social advantage.

Locating the Barrier on the Slope

To see what this barrier would and would not allow, let us pretend that there is one gene for the trait called height, and that different variants lead to people having different heights (in actuality, height is the result of many genes, interacting with each other and with the environment). A male embryo with the rare variant that leads to a four-foot, eight-inch adult could be modified by installing a prevalent variant, which would result in ordinary height. This would remove this person's disadvantage but not give that child social advantage over others. However, a male embryo with the prevalent variant that leads to a five-foot, ten-inch adult could not be changed to have the rare and social advantage producing a six-foot, six-inch variant because it would not be prevalent.

Similarly, let us pretend that there is one gene for "intelligence," and the embryo you have produced has a rare variant that is associated with being much less intelligent. With this barrier in place, you could modify this gene using a prevalent variant, which would presumably be associated with the trait of average intelligence. You could not install the social advantage producing the Einstein variant, which would presumably be rare and not prevalent.

Obviously this barrier is below the fallen germline barrier because it separates germline modifications into acceptable and unacceptable. It is also below the explicit NASEM barrier of "serious disease" because any trait can be modified, and not only those thought to be "serious diseases." This is below where many people want to be on the slope. For example, the critique of HGE from the disabilities studies perspective points out that people who have the rare variant that leads to traits like achondroplasia (a type of dwarfism) do not necessarily think they have a disease and do not necessarily think their children should be "upgraded" to a prevalent variant.[56] With the prevalent variant barrier in place, people *could* modify their children to give them the most prevalent variant for that gene that leads to achondroplasia. Therefore, satisfying the concerns of the disabilities studies community requires either the old germline barrier or a version of a disease barrier that will not hold. This is the ground on the slope that must be sacrificed for the dominance of autonomy felling the germline barrier and for not being able to create a defensible disease/enhancement barrier.

Design Strength of the Barrier

Again, this book is ultimately based in sociology, not ethics, so I am not writing about whether this barrier is the correct ethical position in the HGE debate. I will leave that to others. I am interested in whether the barrier would continue to stand, and a strong barrier resists slippery slope mechanisms. The main design strength of this barrier is that the terrain on both sides ("prevalence" of the variant in the population) is defined by variant *frequency* not the cultural definitions of "normal" or "disease" that have made the disease/enhancement barrier so unstable. Prevalence becomes a number and is therefore not subject to the similarity vagueness that knocked over both the disease/enhancement and the somatic/germline barriers. The NASEM does not define "prevalent," or associate a percentage with the term, but says that it means common and not rare.[57] Prevalent is defined in the dictionary as "widespread in a particular area at a particular time."[58]

The reader will immediately recognize that this prevalent variant barrier trades the advantage of no similarity vagueness for severe susceptibility to continuity vagueness with the word "prevalent" (e.g., the difference between 23 and 24). One endpoint of the possible meanings that would stretch our common understanding of "prevalent" is "one or more" humans in the population. If "prevalent" is interpreted as "at least one human," parents could use any human gene variant on the planet, treating the collective human genome like a catalog to select from. For example, if there were one unusual human who is genetically immune to HIV, a wealthy person could add this gene to their sperm and give their child social advantage. Moreover, if Einstein had his genome sequenced, and there was one gene for "top intelligence," then this variant held by this one human at one point in time could be installed in people's children. This would then allow for the social advantage applications that would violate the value of fairness (advantageous variants will be rare).

Of course, "prevalent" is not totally vague, and it does not mean one or more. To perhaps stretch my metaphor beyond all usefulness, this barrier would fit better with the value of justice or fairness as people interpret "prevalent" to mean larger and larger frequencies. This would make the variants you can select more common and thus presumably less capable of providing advantage.

How else could people interpret prevalent? The synonyms include common, commonplace, everyday, extensive, frequent, new, normal, popular, prevailing, rampant, rife, ubiquitous, universal.[59] I would say that all of these words mean "half," which would only allow the installation of variants so common that they would not give advantage. But, the fact that you (the reader) are debating my interpretation of these words suggests that this barrier will be very susceptible to continuity vagueness.[60] It seems inevitable that someone will say that a variant held by 49% of the population is prevalent and should be upslope of the barrier, which would begin the slide to 48%, 47%, and down. There are a number of institutionalized meanings of "prevalent" from elsewhere in genetic science, but all reflect a large enough percentage that this could potentially allow for selecting a variant that would give advantage and thus could not define a barrier supported by the value of justice or fairness.[61]

An Amendment to the Design of the Prevalent Variant Barrier

The prevalent variant barrier that can be built from the NASEM report is susceptible to continuity vagueness. This problem can be fixed with one tweak—changing the variant that can be installed from "versions that are prevalent in the population" to "the version that is most prevalent in the population." This removes the continuity vagueness because there can be only one variant that is "the most."

Therefore, people could modify their children to give them the most prevalent variant of the gene. Again, this allows for the treatment of rare monogenic diseases where rare variants lead to disease and the most prevalent variant leads to health. Again, and critically, being normal or average does not generate social advantage. This barrier is therefore perfectly supported by the value of fairness as defined above.

Of course, "most prevalent" is in reference to a population, and population could mean the population of the entire world, of the region (e.g. Africa), the nation-state, or another socially defined group. Different populations may have different variants that are the most prevalent. Genes interact with the environment to produce an outcome, so there are some environments where a variant could be good for you and others where it could be bad for you. The classic case is that being heterozygous for sickle-cell makes someone resistant to malaria, which is useful in Africa where malaria is endemic, but not in the United States, where malaria is rare.[62]

Absent the emergence of a functional world government, there are reasons to think that the overarching value of justice or fairness that justifies this barrier will also set the population at the nation (or supra-nation in the case of the European Union). With perhaps a few noble exceptions, nations do not think of themselves as primarily responsible to the world. They are not going to be concerned that they have social advantage over the people of other nations. Rather, they are concerned at best that people within the nation's boundaries do not have social advantage over each other.

That said, it is conceivable that someone could gain social advantage in one country by using the most common variant from the population of another country. It would be difficult to come up with a moral distinction between populations so this barrier would in principle be subject to similarity vagueness. However, people would have a hard time "population shopping" to identify a most prevalent variant in a national population that gives social advantage in their own nation because, while rare variants are concentrated in region of origin, the most common ones are not—as we would expect from evolutionary theory.[63] If it is the rare variants that can produce traits that give advantage, a rare variant in one national context is not going to be the most prevalent variant in another.

In sum, the most prevalent variant barrier is not susceptible to slippery slope processes, and the acceptable upslope region would be any modification that uses the most prevalent variant of a gene—a change that would level the playing field for the disadvantaged while not generating a trait that gives social advantage for the already ordinary. The downslope, unacceptable side of the barrier would be using a variant that is not the most prevalent, which could generate social advantage.

Future Polygenic Selection and Modification

The NASEM report advocated knocking over the germline barrier due to the slipperiness between germline selection (PGD) and modification (editing). PGD has only been available for single-gene traits until very recently, and since the committee thought it would be many years before CRISPR could be used to safely modify even a single gene in a gamete or embryo, there was no point in creating a more generalizable scheme for future decades.

However, the most prevalent variant barrier will be structurally sound if HGE only remains possible for monogenic traits, and I want to try to anticipate the day when multiple genes for polygenic traits can be selected or modified. The ability to select or modify polygenic traits would topple the most prevalent variant barrier because in a polygenic context, there is no most prevalent variant. Without another barrier in place before the advent of polygenic selection or modification, all acts below would be acceptable.

Before continuing, I should note that readers who are very familiar with the scientific literature on CRISPR may conclude that at this point we can rely upon our old friend in these debates—the scientific reality barrier. In the present case, the scientific reality barrier would say that the debate will never fall below the barrier because it will always be impossible to select an embryo with, or to modify, for example, the 100 or more genes that combine to produce a polygenic trait.

We have long looked to the scientific reality barrier to save us from ourselves. Reflecting on the eugenics movement, R. S. Morison long ago wrote that "the thing that has saved man from his limited visions in the past has been the difficulty of devising suitable means for reaching them."[64] However, it is not wise to bet that our knowledge will remain limited, as it was only a few years ago when safe germline modification appeared to be impossible. It is important to identify barriers that are not going to be approached any time soon, so that they can become institutionalized, in the same way that the germline barrier became institutionalized at a time when nobody had any idea how germline HGE might actually be attempted.

To refresh our memories about human genetics, there are a number of monogenic traits—traits caused by a variant of one gene—that are largely considered to be diseases, and that one gene is critical for some cellular function, the absence of which leads to dysfunction. For example, sickle-cell anemia, where the red blood cell is misshapen and then does not work properly, is the result of one nucleotide substitution (i.e., A to a T) in one gene. These are the classic "genetic diseases" that the HGE debate has long focused on. These single-gene traits are what the most prevalent variant barrier can address.

But, much contemporary research is not about these classic genetic diseases but about variants or combinations of variants that result in increased propensity to diseases not usually thought of as genetic, like heart disease. These propensities are largely polygenic—the product of many genes acting together. Most importantly, the traits that would give the most social advantage are most certainly polygenic. For example, a recent study concluded that the "genetic architecture for human height" is "characterized by a very large but finite number (thousands) of causal variants."[65] Unlike a single-gene disease that you typically have or do not have, polygenic traits are more likely continuous. For example, the trait of height is measured on a continuum.

How do we make the connection between multiple genes (the genotype) and the resulting human (the phenotype)? It used to be hugely expensive to determine where causal gene variants were on the human genome, taking years of study to find one variant associated with disease. Now, gene mapping is in orders of magnitude less expensive, and recent years have seen the emergence of genome-wide association studies (GWAS). This is a "big data" approach to genetics, where a huge amount of genetic data about millions of people is examined, and correlations are identified between particular gene variants and traits. These associations will get more and more precise as more people have themselves sequenced, through companies like 23andMe, and those data are used in these studies.[66]

GWAS have made it clear that there are not only a huge number of variants that seem to make no difference to how a body functions but that also the vast majority of observable traits are due to dozens or hundreds of genes combining or interacting in some unknown way, with each variant contributing a tiny bit of predictive value to the outcome.[67] But, you can combine the data about all of these genes from a million different people into a score, called a polygenic risk score. Although you do not know how the genes lead to bodily effects, mining through the collected genomic data may show, for example, that there is a particular combination of 100 alleles that results in a 10% increased risk of that person having schizophrenia.

If you can do this for disease, you can do it for any trait for which you have a measure in the people whose genes are being tested. For example, an article published in *Science* magazine was able to predict a person's educational outcome by examining a polygenic score without any notion of how the connection between genotype and educational outcomes operated. It is predicted that better measures are just a matter of obtaining more data and that more and more traits will have scores. Dalton Conley, who is both a sociologist and a geneticist, predicts that "it will not be long before you can hire companies to take your genetic data and create polygenic scores for anything—from educational attainment to BMI to entrepreneurship proclivity to risk of depression."[68]

People could use these polygenic scores for HGE in two ways: selection (through PGD), and modification through gene editing. Up until now, PGD has allowed people to select for single-gene traits in embryos, and single-gene variants are by and large associated with what are consensually thought to be diseases. It looks like soon whole genome sequencing of embryos will be inexpensive and fast enough to be used to select an embryo to implant based on the hundreds of genes in a polygenic risk score. Conley and Fletcher predict a future where a whole genome sequence of an embryo can estimate "eye color, predicted height, predicted BMI, predicted IQ, and even predicted income of the embryo."[69]

While this would first come to pass for selection via PGD, it could eventually be used for germline modification with gene editing. If one knows how to identify the variants to select embryos, one can identify the variants to modify in embryos or gametes. However, the modification version of HGE based on polygenic scores is truly far off in the future. GWAS only identify the genomic *regions* that hold genetic determinants of complex traits, and it is these markers that are associated with outcomes. For selection via PGD you just need this broader measure. What is needed for actual modification is to identify the actual causal variant that leads to the trait through what is called "fine-mapping."[70] It will be quite a while before the causal variants in a polygenic risk score for educational attainment are identified so that they could by modified. Therefore, selection with PGD will be possible first and will drive developments in the debate.

Designing the Median Trait Barrier

To account for the advent of polygenic scores in HGE, the most prevalent variant barrier could be redesigned to be the median trait barrier. "Median" means the middle person in a population ordered along some scale. Therefore, with this barrier a child could be selected through PGD or modified with gene editing to give them the median value of a trait in a population. Above the barrier would be selecting or modifying any of the variants in the polygenic scale to produce a child that is at the median or below of the trait measured by the scale. Below the barrier, and thus not allowed, is producing a child that is above the median of the trait in question.[71] I will explain below how this barrier is supported by the value of justice or fairness, but for now I will simply note that the median is more consistent with the value of fairness than other measures of central tendency in a distribution such as mean or mode.[72]

For example, for the trait of polygenic educational attainment, parents could give their children the combination of variants that match the human in the database who is closest to the median of that score or select the embryo that was closest to the median. That would "level the playing field," but not allow parents to produce social advantage by selecting a combination above the median.

Contemporary reform eugenicists and transhumanists will be happy that the use of this barrier would slowly produce the improvement of the human species—just without genetically generated social inequality along the way. That is, if these practices became widespread, the people who would otherwise produce a baby at the bottom of a polygenic IQ scale would "upgrade" their children to the median. This would make the overall population have slightly higher IQs, moving the median up a notch, and the next generation of children would be upgrading to a median that is a bit higher, resulting in the slow ratcheting up of median IQ in the population. For those who are only using the values of autonomy, beneficence, and justice (fairness), this is a great outcome. But, this is one of the many reasons why many participants in the HGE debate using other values would want to defend the germline barrier.

Challenges to the Median Trait Barrier

This barrier, located at the human in the data closest to the median of the scale, does not suffer from continuity vagueness. There may be more than one such human, thus many combinations that produce the scale value, but there will only be one scale value that is "closest to the median."

This barrier is only fairly resistant to similarity vagueness. Previous barriers based purely on social convention, like the disease/enhancement barrier, have been susceptible to similarity vagueness compared to those with an anchor in how human biology works, like the somatic/germline barrier (e.g., cells modified in the spinal cord are not passed on to one's children). Once we get into the polygenic context, this becomes undeniably social as the traits that define these scales are socially defined. For example, what is education, why do we care about it, and how do we measure it? While it is possible to define this barrier objectively, I do not consider that definition further because whether it would be consistent with the value of fairness is unclear from our present knowledge of genetics.[73]

This problem of non-objective definitions is mitigated when we realize that the only value in use at this point is fairness, and therefore the only scales that matter are those designed to measure socially agreed upon traits that confer advantage or disadvantage. If someone comes up with a scale irrelevant to advantage, like the "number of eyelashes scale," such a selection or modification is irrelevant to the values in the debate. Moreover, anyone who spends the money to create a scale is going to have "bad" at one end and "good" at the other. However, to actually enact any of the barriers in this book would require a regulator of some sort (as we have with the NIH deciding what experiments are allowed with NIH funding). The regulator would have to keep a list of all established polygenic risk scores, and people could make their children like the median of any of them—be it height, educational attainment, or toe length.

There is also the problem that desirable traits are not unitary. There is not some objective measure of "intelligence," but people would say there is emotional intelligence, interpersonal intelligence, analytic intelligence, and so on. However, this is merely a problem of labeling the polygenic scores. Imagine, and it is not hard to imagine, someone coming up with a polygenic risk score that predicts the results of an IQ test. Who knows how to actually describe whatever it is that IQ is measuring? But that does not matter, and whether a high IQ actually would give advantage does not ultimately matter. What matters is that everyone thinks that it does.

Values That Support the Median Trait Barrier

All I say about values and the median trait barrier in this section also holds for the simpler most prevalent trait barrier. The primary value that justifies these barriers is a particular version of justice or fairness. Justice or fairness is a deeply entrenched value in at least the West, and as one of the core American myths, all public figures have to claim to be forwarding this value. For example, as of this writing, the Republican Party headed by President Trump claims that its primary constituency is not being treated *fairly* by the elites, and the United States is not being treated *fairly* by other countries. Democrats claim that Republican tax cuts for the rich are unfair to the middle class. Regardless of the truth of these political claims, I think it is clear that fairness is part of the deep discursive structure of American society, even when it is being breached.

Of course, this is a particular version of fairness. Traditionally, for ordinary citizens, fairness does not require creating equal outcomes for people but rather creating equal opportunity to succeed "by your own bootstraps." Reflecting the power of this value, in one study, 98% of the public agrees that "everyone in America should have equal opportunities to get ahead" and that 98% of the public agrees that "children should have equal educational opportunities."[74] Few would support the government buying everyone a house, but there would be much more support for government paying for education that allows people to use their own initiative to compete for jobs that would allow them to buy a house.

Raising the genetically disadvantaged to the median—neither advantaged nor disadvantaged—would be seen by many as creating a fair playing field for people to work hard within. However, allowing the rich to solidify social advantage for their children through genetics would be unfair, like allowing someone who inherited wealth from their parents to buy the beach so that only they can use it.

Whether the public has this exact politically liberal view of fairness is not as important for what happens with HGE policy as whether the academics who participate in the HGE debate have this view. My assessment is that they do. Nearly everyone in this debate has a doctoral degree and is a professor, and this tiny subpopulation of the United States is disproportionately liberal, particularly at the elite institutions from which the HGE debate draws its participants.[75] Moreover, the most influential participants in these debates are from the profession of bioethics, which strongly advocates the value of justice or fairness, which further strengthens this barrier. For example, justice was the only explicitly identified value by the NASEM summit organizing committee beyond beneficence and non-maleficence.[76]

The notion of justice as used by bioethicists is greatly indebted to the views of political theorist John Rawls. For example, the NASEM committee relied upon what is probably the most influential text by bioethicists in the 21st-century HGE debate, written by Buchanan, Brock, Daniels, and Wikler, which is heavily indebted to Rawls.[77] These authors conclude that justice requires leveling the playing field in society, which gives everyone an opportunity to compete fairly. They extend this notion of justice to genetic inheritance, and they therefore advocate the creation of a level playing field in natural biological differences between people because the genetic traits you were born with are not chosen by you and out of your control. They want a "genetic decent minimum" for all people.[78]

They are agnostic about whether HGE should be limited to diseases, writing that to decide we "would have to know whether there are genetic conditions that significantly limit opportunities but that do not constitute diseases and that we will be able to prevent or ameliorate though the application of genetic science."[79] Their advocacy of the value of justice fits well with the median trait barrier in that potential parents could identify traits for which their potential children lack a genetically decent minimum and bring them up to the median.

It is not only those who want to proceed with germline modification who support the value of justice but skeptics as well. Indeed, I would argue that many liberals who defend the germline barrier do so because they do not see any viable barriers between the germline and the dystopian bottom of the slope of genetically derived social stratification. For example, the left-oriented critics at the Center for Genetics and Society would think—probably quite rightly—that the somatic/germline barrier is less susceptible to slippery slope processes, and they would think that the median trait barrier is further downslope toward their dystopia. But, the value of "unfairness" below the median trait barrier accurately depicts much of what concerns this group about the bottom of the slope. As I quoted earlier, they declared in a statement that "permitting germline intervention for any intended purpose would open the door to an era of high-tech consumer eugenics in which affluent parents seek to choose socially preferred qualities for their children. At a time when economic inequality is surging worldwide, heritable genetic modification could inscribe new forms of inequality and discrimination onto the human genome."[80] This dystopia could not happen with the median trait barrier—if people are only allowed to modify their children to make them ordinary (i.e., at the median). That is, if the barrier holds.

Values That Imperil the Median Trait Barrier

A Different Interpretation of the Value of Justice

At this point in the book I begin engaging some arguments that have not yet appeared but that I anticipate. The strong design of the barrier—its limited susceptibility to slippery slope processes—means that toppling the barrier requires the more difficult task of redefining the terrain by changing the values. The most immediate threat to this barrier will come from a different interpretation of what the value of liberal justice requires, which would redefine downslope acts as fair. I repeat the words from the NASEM report, with my emphasis:

> [A] problematic enhancement is one that confers a social advantage beyond that which an individual possesses by fate or through personal effort, *and that does not benefit the rest of society in any way* or undermines the implicit goals of a competition. Using equality of opportunity *and societally useful inequality* as guides may help distinguish those forms of enhancement that might generally be tolerated (assuming the risks are proportional to the benefits) from those that would be more controversial.[81]

That is, by some interpretations, some individual inequality is good for the entire society. For example, doctors get paid more than street sweepers, which encourages the most talented people to become doctors, which benefits everyone. Some will undoubtedly argue that using HGE to make some children be above the median on a scale will ultimately benefit everyone.

Let us imagine a modification that would be below the median trait barrier and not allowed. Eric Lander of the Broad Institute at Harvard identifies some applications that many scientists will want to do:

> Another potential application is reducing the risk of common diseases, such as heart disease, cancer, diabetes, and multiple sclerosis. The heritable influence on disease risk is polygenic, shaped by variants in dozens to hundreds of genes. Common variants tend to make only modest contributions (for example, reducing risk from 10% to 9.5%); rare variants sometimes have larger effects, including a few for which heterozygosity provides significant protection against disease. Some observers might propose reshaping the human gene pool by endowing all children with many naturally occurring "protective" variants.[82]

Using the median trait barrier, such gene variants could not be installed in children because they would produce a child that is above the median on the trait of protection against disease. (This would also not be allowed with the germline, prevalent variant or most prevalent variant barriers.)

Whether making your child have a reduced risk for heart disease benefits the rest of society depends on how you think society works, and I will preview the arguments that will be made so the reader can decide how likely it is that a different definition of fairness will redefine the acts just below the median trait barrier. The first argument for this alternative version of fairness, and thus moving beyond the barrier, is that the technology would benefit everyone because everyone would have access to it. This would probably be the position of some European welfare states with national healthcare systems. However, there are only six countries in Europe that offer complete coverage of costs for assisted reproductive technologies,[83] so it seems unlikely that even all of Europe would make genetic modification available to all. Perhaps in Denmark.

In countries without universal access to healthcare, like the United States, this technology would only be available to the wealthy. But, I have heard many scientists say that they presume that the medical technology developed for the benefit of rich people eventually trickles down to poor people. By extension, the argument would be that experimenting on rich people who want to give their child advantages generates knowledge that can be spread to everyone.

The counter-argument is that, at least in the United States, the most generous interpretation is that technology trickles down exceedingly slowly. For example, at the time of a 2009 study, 46 million Americans lacked health coverage, which is associated with an extra 44,789 deaths per year.[84] This is a case of technologies developed long ago not trickling down to the poor—for however long we have had penicillin, X-ray machines, and other basic technologies.

Moreover, critics will continue, the wealthy will hoard this technology to give their children and future descendants social advantage. Something you have only allows for advantage if not everyone has it (what economists consider a type of positional good).[85] A college degree has relative value because only a third of those in the United States have one. Moreover, few Harvard alumni are clamoring to remove the advantage their children have in the Harvard admissions process. Therefore, it will be in the interest of the wealthy to not make HGE broadly available.

However, the main argument against the trickle-down theory is that being modified will become part of one's identity. This is unique to this technology because germline HGE is different from an X-ray machine, a medicine, or even somatic HGE. No one thinks their identity is based on their usage of X-ray machines, a particular medicine, or even that they had the cells in their eyes genetically modified to fight eye disease. In contrast, germline HGE is about reproductive genetics, and genes produce a connection—at least in people's minds—between parents, their children, their grandchildren, and all of their progeny. Evidence for this is in the extreme concern among the proponents of HGE for genetic relatedness between parents and their children. Given how Americans think about genetic relatedness, it seems possible that being genetically enhanced will become part of the upper-class identity. Even if genes do not have the strong effect on outcomes that people presume, it will not matter. All that matters for "parallel populations" to develop is for people to *think* there is a strong effect.[86]

In this scenario, the genetically modified or selected will be encouraged to only mate with others with the same improved trait. More inequality would be produced as the upper classes can prevent diseases genetically and the upper classes will then conclude that the problem of diseases is solved and that funding should be cut for treatment of disease for ordinary people through conventional medicine.

So, expect that those who want to cross the median trait barrier to argue that the benefits will eventually accrue to the entire society, and that is the better interpretation of the value of justice or fairness. Those who want to defend the median trait barrier will have to claim that this is based on a faulty notion of how the society works *and* argue that concerns about fairness trump the health of the rich children who could have been modified.

The Value of Beneficence Threatens ... Again

To be clear, if scientists find a way to take a rare variant that would fight disease and put it in the genome of an embryo, this would install an above-the-median trait in an average person. If the median trait barrier were to hold, then the barrier would cause people to die from disease when we have the ability to save them. Would we really deny saving the lives of rich people's children just because doing so would create more inequality, especially in a world where access to medical care and all other life opportunities are already structured by inequality?

This raises again the value of beneficence, and we can anticipate that if such disease-resistant genes are found, people in the debate will advocate having beneficence justify acts below the median trait barrier. The argument will be that everything said by advocates of justice or fairness is true about "non-disease" traits like intelligence, but that fighting "diseases" should be allowed due to the value of beneficence. Therefore, the argument continues, the median trait barrier should be removed, and we should rely upon a new disease/enhancement barrier which would be rebuilt further down the slope.

This new location on the slope would mean that people could be made above the median in health-related traits, but not for non-health-related traits like educational attainment. But, as I will show in the next chapter, a new disease barrier cannot hold. To defend the median trait barrier, participants in the debate will have to argue that justice or fairness is the more important value, and that not allowing the already advantaged to further their advantage with disease-resistant children is more important than the health of those wealthy children. In my view, this argument is unlikely to succeed.

The Value of Autonomy Threatens ... Again

Another threat to the median trait barrier will be the value of autonomy, based on the argument that anything concerning creating a baby should be the autonomous choice of parents. The development of reproductive medicine more generally, with its ethos of patient autonomy within broad limits, as well as the pro-choice movement in the abortion debate, have created a situation where HGE is increasingly coming to be thought of as a free, autonomous choice, much like the decision to have an abortion.

For example, Julian Savulescu, in an influential paper much commented upon by others, argues for near total autonomy.

Indeed, parents are morally obligated to select embryos or fetuses that are "most likely to have the best life" be it because they are free of genes that cause disease or because they are enhanced. Parents can choose because in liberal democracies "we should allow couples to make their own decisions about which child to have," which includes being "allowed to select a child with disability, if they have a good reason."[87]

Similarly, John Robertson—arguably the father of the concept of reproductive liberty in bioethics—identifies a perspective he did not necessarily agree with, called the "radical liberty" perspective, where "no limits can appropriately be placed on what they do before the birth of a child," allowing people to "select, screen, alter, engineer, or clone offspring as they choose." He writes that this position "hovers in the background and casts a shadow over many official, scholarly, and popular accounts of reproductive issues," even though most advocates of autonomy endorse some limits on freedom, such as not being able to intentionally hurt your offspring.[88] Those Robertson saw as "hovering in the background" should be expected to strongly emerge if the median trait barrier is built.

The Barriers Examined in This Chapter

This chapter began in 2015 when it became known that a team of Chinese scientists were experimenting with the germline modification of human embryos using the new technique called CRISPR. While most groups who were writing at the time reiterated the somatic/germline barrier as the moral limit, this was primarily justified by non-maleficence (safety), and the increased safety of CRISPR would remove the justification for this barrier. A group of American scientists and bioethicists argued for cautiously taking down the germline barrier, which eventually resulted in a committee of the NASEM that produced a report advocating removal of the germline barrier, the placement of a number of conditions that had to be met before we do so, and ultimately relying upon a barrier downslope located at "serious disease."

This conclusion logically flowed from the bioethical discourse in the debate up to this point. What would have been needed for the barrier to remain standing was for someone to have earlier justified and popularized a barrier between germline selection (PGD) and germline modification (gene editing). Lacking such a barrier, the locations on both sides of the traditional somatic/germline barrier suffered from extreme similarity vagueness, and the barrier no longer made any sense.

I examined the "explicit" barrier proposed by the NASEM committee at serious disease, created for use only in the immediate future, and concluded that it will not hold if the technology develops further, as it is susceptible to an extreme similarity vagueness slippery slope process even if no values change in the debate. Again, how would *you* define a disease? A more long-term barrier can be created from the value of justice or fairness articulated in the report and can be set at "the most prevalent variant." This will hold as long as scientists' abilities are limited to monogenic selection or modification.

To create a barrier useful in a future era of polygenic HGE, I identified a barrier located at the "median trait." While supported by an interpretation of the value of justice or fairness, and resistant to slippery slope processes, this barrier could still be felled by other values becoming dominant. As has been the case in this debate for decades, the value of autonomy would make this and other barriers moot as people used their autonomous decision-making to decide the genetic qualities of their children. A conception of justice or fairness where inequality is allowed if it benefits the society in general would be argued to remove the barrier or push it far downslope. Finally, given that the median trait barrier would not allow creating children unusually resistant to disease, the barrier would remain vulnerable to the value of beneficence. As we can see, it is difficult to justify barriers on the slope below the traditional disease/enhancement and somatic/germline barriers using the dominant values in the contemporary debate.

4
Possible Barriers Further Down the Slope

In the previous chapter, I identified the median trait barrier, which is the strongest barrier I could identify immediately below the germline barrier. Unlike the old disease/enhancement barrier, which suffered from extreme similarity vagueness, the new barrier is fairly structurally sound. The primary threat to this barrier is by a change in the dominant values in the debate, which is a more difficult way to topple a barrier than showing continuity or similarity vagueness.

But, I can already see that many people in the debate are not going to accept the median trait barrier when it becomes possible to install unusual variants that protect against disease. Indeed, that is what the Chinese scientist was trying to do in 2018 by attempting to create the twin girls who are resistant to HIV.[1] People will correctly identify that the median trait barrier limits the application of the value of beneficence.

In this chapter I identify possible barriers further downslope from the median trait barrier, roughly presented in the order of higher to lower on the slope. I evaluate two types of barriers: Those that currently have prominent support in the debate (although never described using the concepts of slippery slopes and barriers), and those that are not discussed but would be structurally sound according to slippery slope theory. The general conclusion is that the barriers get weaker as the debate moves downslope from the old germline barrier, but there are strong barriers defending territory near the dystopian bottom of the slope.

It is important to recognize that for the median trait barrier, the values of those who Jonathan Haidt calls religious and cultural conservatives are no longer influential. In the earlier debate, before the turn of the century, they were the ones with values such as avoiding hubris, "not violating nature," or "not violating God's will." They were concerned with what humans "truly are," and thus they were (and some still are) concerned with issues such as whether in a genetically focused society people would be considered somewhat more like objects and therefore dehumanized.

The barriers downslope of the germline barrier, including the median trait barrier, are largely built on what Haidt would call liberal values—autonomy, beneficence, non-maleficence, and justice (fairness). This is not surprising given that the people with religious perspectives largely left this debate decades ago.[2] That said, I will identify a few barriers, albeit near the bottom of the slope, where these old values may be used.

This chapter, by necessity, is even more speculative than the previous one. At least in the previous chapter, I largely referred to technology that is plausibly envisioned in the near future and ethical proposals already made. Here I speculate about technological developments and describe various imagined futures based on how societies operate. As I discussed in the Introduction, I think that the benefits of speculating outweigh the risks.

So, I start with a general scenario. Imagine a future where technology improves, embryos are selected on polygenic traits, and more than one gene in a gamete or embryo can be edited. The US government tries to promulgate regulations based on the NASEM report. When writing these regulations, the government gives up on the explicit NASEM barrier because they find the terrain around "severe disease" hopelessly susceptible to similarity vagueness. Still not wanting a situation where people can modify their children in any way they want, they adopt the median trait barrier as the only available solution less susceptible to slippery slope processes and for which they have justification using the values prominent in the debate. Policy is promulgated, and couples who would produce embryos likely to be below the median in some trait start using HGE to provide their children a combination of variants that leads to the median of that trait—making their offspring that otherwise would have been at a disadvantage ordinary. If this future were to come to pass, a number of forces could threaten this barrier, as described in the previous chapter. If the median trait barrier falls, what could be built further below it?

The Goals of Medicine Barrier

Let me start with a barrier that has previously been proposed, and will be again if it becomes possible to not only correct genetic diseases but to also give health advantages via HGE. This barrier originally emerged when the old disease/enhancement barrier could not address enhancements that fight disease. For example, in 1998 Erik Parens identified one of the (many) problems with the disease/enhancement barrier, which was that "both interventions aimed at treating disease and ones aimed at enhancing human performance are improvements."[3] Similarly, Eric Juengst observed in 1997 that maintaining the disease/enhancement barrier at the "normal functioning" standard would preclude "preventive interventions that act by strengthening an organism's normal defenses against disease."[4] This critique of the traditional disease/enhancement barrier is best understood by trying to square the current use of vaccinations with the barrier. Vaccinations create enhanced humans. How can we accept vaccine-enhanced humans but not genetic-enhanced humans?

Given the inability to define the disease/enhancement barrier, another school of thought in the HGE debate simply outsourced the definition to the medical profession. This is the "professional domain account" of disease, which "appeals to the health professions' conventional vision of their proper domain. Accordingly, 'treatments' are any interventions that the professional standards of care endorse, while 'enhancements' are any interventions that the professions declare to be beyond their purview."[5] This creates the goals of medicine barrier, where above the barrier is what the profession of medicine considers to be a disease and below the barrier is what the profession does not consider to be a disease.

Weaknesses of the Goals of Medicine Barrier

The goals of medicine barrier would be supported by the same values that support the original disease/enhancement barrier. The basic question is: How does the medical profession define its goals? The problem with this barrier becomes immediately apparent. As Juengst and Moseley write, "given the health professions' philosophical pluralism and political autonomy, their own conventions seem to provide no principled way to exclude new interventions from their domain."[6] For example, engaging in surgery to allow someone to achieve their aesthetic goals is just as much a goal of medicine as engaging in surgery to repair broken bones. If cosmetic surgery is one of the goals of medicine, then, analogously, using HGE to install a rare gene variant that makes your child more attractive will be a goal of medicine.

Indeed, the goals of medicine can be widely interpreted. The limitless expansion of the medical domain is classically indicated by the 1947 statement of the World Health Organization that defined health as "a state of complete physical, mental and social wellbeing and not merely the absence of disease." An example of how this barrier would be located near the bottom of the slope, improving "social wellbeing" would allow making children above the median in terms of intelligence.

Even if we could initially place this barrier at a particular location on the slope, it would slip downhill due to medicine's infamous jurisdictional expansion, termed "medicalization" by sociologists. Medicalization is the process by which normal body functions and/or social problems come to be defined as "medical problems." Most famously, childbirth is a normal, natural process that used to be under the jurisdiction of midwives but is now in the medical domain. Similarly, pharmaceutical companies seem to invent diseases in order to sell the drugs they have developed to treat them. Alcoholism, drug addition, sexuality, and many more phenomena have come under the jurisdiction of the medical profession.[7] Indeed, the primary motivation for the NASEM taking down the somatic/germline barrier stemmed from defining the parental desire to have genetically related children as a medical problem to be addressed by doctors. In Juengst and Moseley's somewhat more generous description of creating a barrier defined at the goals of medicine, they state that this barrier is challenged by "biomedicine's infamous nosological elasticity. It is not hard to coin new maladies for the purpose of justifying the use of enhancement interventions."[8]

Therefore, while it would slow slipping down the slope, as the culture of the medical profession does not change overnight, the goals of medicine barrier would not function as a barrier in any sense of stopping further sliding. To perhaps overuse the metaphor, this barrier would just slowly slip down the slope as the goals of medicine changed. Moreover, the trajectory of medicine is away from limiting its domain to disease and suffering and toward satisfying the bodily desires of patients—such as by offering enhancements.

The Family Genes Barrier

The family genes barrier is located at modifying the genome of a child to a set of genes that the couple could have possibly produced with sexual reproduction. This barrier is a response to a possible technological development, and has at present no proponents of which I am aware. In the last chapter I described the possibility of whole genome sequencing (WGS) of embryos in a dish and giving each embryo a polygenic score. For example, each embryo could be ranked by future predicted susceptibility to heart disease or educational attainment. This *selection* through WGS-PGD could be constrained by the median trait barrier.

If this version of PGD is all that is available, there will be only mild pressure to take down the median trait barrier. Whereas we can imagine couples using this technology to avoid having children with traits that are widely accepted as diseases, especially if the parents were using IVF anyway, few would try to use WGS-PGD to give their children a trait that would offer an advantage. The reason is that they could only select from at best a dozen embryos that they had produced, and the odds of one of them having a rare advantageous genetic trait are low. Moreover, people will not use PGD lightly. Extracting eggs from a woman is a very unpleasant and somewhat risky procedure, and the odds of that one selected embryo actually becoming a baby in the end are not that high.

As an example, let us say that polygenic risk scores are distributed in a bell shape, with most of the scores clustering around the median and a few outliers far from the median. Height would have such a distribution, with most people in the middle and with long tails—with the four-foot, six-inch tall men on the left end and the seven-feet tall men on the right. If you have 12 embryos to select from, odds are that most or all of them will be clustered in the middle. Would you do any better if you relied on chance and made a baby the old-fashioned way? Or, let us say that a couple would produce one Einstein in 500 babies. For people to want to use WGS-PGD for intelligence, they need to be able to produce that one Einstein baby, and that is unlikely to happen with only a dozen embryos to choose from.

The technological improvement that would generate pressure against the median trait barrier would be if people could create artificial embryos through induced pluripotent stem cells derived from their skin, avoiding egg harvesting, which is the most dangerous and unpleasant part of IVF. This is apparently so close to becoming reality that some scientists are already calling for a ban on the use of this technology for reproductive purposes.[9] More importantly, presuming further drops in the price of sequencing the genes of a cell from an embryo, since sperm are plentiful, with artificial embryos a couple could have hundreds of embryos to select from.[10] In this scenario, the odds of creating that Einstein outlier are dramatically greater, increasing the motivation to use this technology.

If this technology were to come to pass there would be pressure to create a meta-level "better baby" score due to extreme information overload. One analyst writes that "several embryos might have numerous traits that are both desirable and undesirable to the parents. Imagine, for example, the genomic analysis of one embryo: female, who would experience early graying and be moderately tall with a heavy build; 65% chance of scoring in the top half of the SAT tests; good chance of above average athletic ability; and likely to be introverted. Now imagine the genomic profile of another embryo: male, who would have male-pattern baldness, medium height, and medium build; 40% chance of scoring in the top half of the SAT tests; likely to have above-average musical ability; and likely to be anxious." How would one decide? Potential parents would probably be presented with an algorithm that sorted through the hundreds of thousands of combinations of traits and created algorithms focused on illness, propensity for athleticism, "life success," and so on.[11] Knowing they could do it, couples would want to have children that maximized a "better baby" score, which by definition means picking a score above the median of the scale.

Depending on some still unknown scientific facts, we would at that point be very close to the liberal version of the bottom of the slope—this is literally the plot line of the movie *Gattaca*, where people select "the best" child possible. How far this is from the bottom depends upon how heritable the traits are that give social advantage. "Intelligence" is partly heritable, but what would make this technology used more would be if there were very high odds of producing an embryo that results in a far more intelligent child than what the parents would typically produce. That is, does a couple, each of whom is 30% below the median of the population in intelligence, only produce embryos with an intelligence centered on the parents' average, but with a distribution reaching down to 60% below the median and up to the median—with a few outliers 20 points higher and lower? Or, is it a flatter distribution with really long tails—and any set of parents will have an Einstein to select? If a couple can really boost the social advantage of their children through this PGD, then there will be great motivation for people in the public bioethical debate to find reasons to take down the median trait barrier. In the last chapter I identified how it could be taken down.

Design of the Family Genes Barrier

If this technology comes to pass, and the median trait barrier does not hold, then people would be able to use selection WGS-PGD to produce a child with any traits that the genomes of the couple can produce. That would then become the de facto location of the barrier, and it would be important to define the barrier in such a way that it could also be used as the barrier for gene modification through gene editing. If there was no barrier, then people could search through the collective genes of humanity looking for variants to use to gain advantage. I repeat the caveat that selecting polygenic traits will be possible long, long before modifying a polygenic trait.

How would we define a barrier that allows the same acts for both selection (WGS-PGD) and modification (gene editing)? Above the family genes barrier would be the traits that the couple could produce in a baby via WGS-PGD, and below would be the traits that the couple could not produce through WGS-PGD.[12] Users of WGS-PGD and modification HGE would then be treated similarly but not exactly the same.[13] If the couple could produce one Einstein out

POSSIBLE BARRIERS FURTHER DOWN THE SLOPE 119

of unlimited embryos, then they could also modify one embryo to give it the traits they could produce. If they could not, they could not modify a germline cell to give it the Einstein trait.

The design of this barrier limits slippery slope processes. There is no risk of continuity vagueness creating a slippery slope process with this barrier. That is, the terrain immediately on either side of the barrier is not associated with a number, so there is no risk that a 22 can justify a 23. There is lower risk of similarity vagueness in that the variants that could possibly be produced in a child by the reproducing couple is not vague.

[margin note: But how to justify this barrier? Won't it seem unfair and arbitrary (see next page)]

However, there would still be the couples who could produce an Einstein because they have the variants that would possibly create the baby genius, but they are also both homozygous for a disease like sickle cell. Every embryo they produce would also have sickle cell, and the family genes barrier would require modification to a genome you could actually produce, resulting in a baby with the disease.

This is the issue that felled the germline barrier in the NASEM report, and it is hard to imagine a barrier being built downslope that takes away applications that have already been allowed. However, with the family genes barrier, you cannot go outside of the family looking for healthy genes if you are both going to pass on sickle cell.

To accommodate the few hundred homozygous couples, as all barriers below the germline barrier have done, the pool of genes to select from needs to be expanded beyond the couple, for all those who want to use the technology. It cannot be restricted to just those with "disease" because by this point in the slope there is no longer a disease/enhancement distinction. The key is to accommodate in a way that does not open the barrier to slippery slope processes and which does not allow the violation of any of the values that support the barrier, such as by allowing people to find genes that are even more socially advantageous.

To accommodate the homozygous couples, the couple could trade any gene variant in their modification plan with a variant from one of the couple's biological parents. Presumably a couple who are both homozygous for sickle cell would have parents where not all four are homozygous. While it is possible that the two homozygous prospective parents themselves have all four parents who are homozygous for the disease, this is approaching the range of statistically impossible.[14]

The purpose of this barrier is to limit the ability to search the globe for advantageous variants—to find the Einstein genes you were incapable of producing on your own. Since the parents of the couple are included in the pool, the couple can do more searching than they could with just themselves, but they could not find unrelated humans on the other side of the planet to mine for advantageous variants.

Adding other family members to the available pool adds a degree of similarity vagueness because it could be asked: Why a parent of the reproducing couple? Why not a sibling? Why not a cousin? Without a principled reason for distinguishing between parent, sibling, and cousin, this barrier on the slippery slope would erode as the list of family members expanded enough to allow people to have a wide search for advantageous gene variants. This barrier probably sounds utterly implausible to anyone who has participated in these debates, but that just shows the difficulty of constructing barriers below the somatic/germline.

Values That Support the Family Genes Barrier

The median trait barrier was supported by the liberal value of justice or fairness. The family genes barrier would be supported by a more conservative version of fairness. The difference is akin to the difference between progressive taxation and a flat tax. In progressive taxation, those who are already advantaged are assessed a higher tax rate, which limits their ability to further solidify their advantage. Those who are economically disadvantaged have a much lower tax rate and can presumably use more of their money to catch up to the median.

Analogously, with the median trait barrier, if a couple who was already "very intelligent" wanted to further improve the "intelligence genes" of their offspring, they could not do so but would have to accept the most prevalent version of the "intelligence genes," which would be "normal." Analogously, a progressive tax of 60% on income over $5 million would not allow these lucky few to further solidify their advantage but rather regress them to the mean. Similarly, the median trait barrier would be "fair" because it too would not allow the already privileged to further solidify their family's genetic advantages. To use another tax policy analogy, the median trait barrier would result in the redistribution of relative (genetic) wealth in society, bringing the top and bottom closer together. That is one version of "fair" and is probably the one endorsed by most people in bioethical debates.

With the conservative version of the value of fairness embodied in a flat tax, the high- and low-income earners pay the same rate, and the family genes barrier is supported by the same version of fairness. A couple who is already genetically advantaged can use what they have to further their advantage by identifying the "best" genetic offspring they could produce. For example, if both parents scored a 50 on the polygenic educational attainment score, they could through some random combination produce a child who was not a 50 but an 80. A couple who is genetically disadvantaged (both scoring a 30) could end up with a child 30 points higher—at 60. There is no redistribution—you work with what you have.

In addition to a conservative version of justice or fairness, this barrier would also actually be supported by a loose interpretation of the values of following nature or God's will. Upslope is creating a child that the couple could have produced through old-fashioned sexual reproduction "the natural way," and PGD just ensures that a particular possible outcome occurs. This interpretation obviously only holds if we ignore all the incredible amount of human intervention along the way. Below the barrier is that which would be impossible in nature—taking a variant from another human and putting it into your child.

This barrier would be supported by another powerful conservative value, which is that children should be genetically related to their parents. The power of this value was articulated in the NASEM report as it served as one justification for taking down the germline barrier. With the family genes barrier, above the barrier would be the variants that the parents *could* produce, and below that which they could not produce. Below would be the equivalent of gestating an unrelated embryo that was portrayed by the NASEM as unacceptable to many prospective parents.

Values That Would Imperil the Family Genes Barrier

The family genes barrier, like the median trait barrier, is threatened by values that justify acts upslope. First, the value of beneficence would justify acts on both sides of the barrier if a disease-fighting enhancement that the couple could not produce could nonetheless be edited into a child. For example, imagine that a rare variant only found in Africans could make Caucasian children resistant to disease. Second, the value of autonomy would justify applications below the barrier on the grounds that parents should have the freedom to install genetic traits from outside their family if they wish.

The Boundary of Humanity Barrier

The boundary of humanity barrier would allow the insertion of a gene variant that any human in history has had, and not allow any gene variant that a human has not had. This barrier, despite being near the bottom of the slope, has a strong design. Imagine that the one human genetically immune to cancer had been discovered. Plans were being made to identify the exact combination of variants held by this person and using HGE to install them into couples who, of course, could not produce this combination of variants on their own. Harvard geneticist Eric Lander's prediction had also come to pass, where "sports-minded parents" wanted "to introduce the overactive erythropoietin gene that conferred high oxygen-carrying ability on a seven-time Olympic medalist in cross country skiing."[15] This would mean all the previous barriers had fallen.

If these practices come to pass, you would think this is the bottom because parents could insert any variant they wanted. However, there are at least a few barriers left. Imagine at this point that there is another animal species where the brain does not build up the plaques that lead to human dementia, and that this trait is due to some genetic sequence unique to that animal. While scientists were still studiously ignoring the biohackers who wanted to install genes in their children from jellyfish that produce a fluorescent green glow, there was a growing constituency for using animal genes to resist human dementia.

Relatedly, moving beyond the human, some scientists are already pondering artificial variants. Lander, in an editorial in the *New England Journal of Medicine*, states that "some scientists might ask: Why limit ourselves to naturally occurring genetic variants? Why not use synthetic biology to write new cellular circuits that, for example, cause cells to commit suicide if they start down the road toward cancer?" He seems to only oppose this because it is not yet safe, writing that "such efforts would be reckless, at least for now."[16]

Therefore, to stop the slide to the absolute bottom, the barrier would be placed at the limit of "human genes," and above the barrier would be any variant that any human has had and below the barrier would be a variant that no human has ever had. I only identify this barrier because it would have a very strong design and is useful for contemplation, but this would be so close to the bottom that I do not think anyone would take the effort to build or defend it.

Design of the Boundary of Humanity Barrier

The boundary of humanity barrier actually exists in the philosophical literature about human enhancement as a specification of the distinction between what philosopher Nicholas Agar called "moderate enhancement" and "radical enhancement." For Agar, moderate enhancements are those that "do not exceed the maximum attainable by any current or past human being." Radical enhancements improve "significant human attributes and abilities to levels that greatly exceed what is currently possible for human beings."[17] The barrier would have no continuity or similarity vagueness, as there is consensus on what life forms are human.

Indeed, the barrier would be strengthened by the distinction between humans and other life forms being so assumed. Anthropologists have concluded that these distinctions are nearly universal across cultures, with societies making distinctions between person, artifact, plant, and animal.[18] In fact, these distinctions are so ubiquitous that many evolutionary psychologists have concluded that making these basic distinctions gave our ancestors competitive advantage in the era of evolutionary adaptation, and therefore the basic biological structure of the human brain has evolved to make these distinctions. Even studies of people with brain damage have shown that people with certain parts of the brain disabled cannot conceive of these differences. For example, one woman cannot recognize anything having to do with animals, but her understanding of plants and artifacts remains unimpaired. This suggests that information about different life forms are processed in different locations in the brain.[19] If humans have an innate desire to distinguish between humans and other life forms, a barrier based on this principle will be difficult to slip across. *[handwritten: What about Neanderthal genes? Bonobo genes? Dolphin genes?]*

Values That Would Support the Boundary of Humanity Barrier

The primary value that could support the boundary of humanity barrier is following nature and/or God, for which there are few proponents in the contemporary debates. The source of the values for this barrier would be, at least for Christians (who are the largest religious groups in the United States), that what it means to be human is central to all of their concerns with biotechnology. To massively overgeneralize, humans are those who are made in the image of God, and animals and other life forms are not. Rearranging genes within the human family does not provide a particular threat to the idea that humans are made in the image of God, and depending on the goal, such rearranging could be seen as a commandment from God to relieve suffering from disease.[20] However, mixing humans with other life forms would not be consistent with the image concept.

Similarly, someone in this debate who places high value on nature would probably argue that humans, as constituted by natural evolution, cannot genetically mix with other life forms: to do so is to violate nature. At that point I would expect the faithful remnant supporting these values in the debate, if such a remnant still exists, to argue against "chimeric humans," "hybrid humans," "synthetic humans," or "GMO humans." Further supporting the barrier would be that moving beyond it would require a debate about what is a human in the newfound chimeric age, and reluctance to have this debate would bolster the barrier. Relatedly, this barrier would be supported by anthropocentrism, the belief that humans are superior to and/or more important than other life forms, which is a cornerstone of both Western religion and philosophy.

Values That Would Imperil the Boundary of Humanity Barrier

While not vulnerable to slippery slope processes, beneficence would justify acts on both sides of the boundary of humanity barrier if a trait no human has ever had would advance human health and reduce suffering. Moreover, as with previous barriers, the value of autonomy would also straddle the barrier, and people would argue that parents had the right to autonomously decide whether to radically enhance their children by using non-human variants.

I can also imagine a value entering the debate that has not yet been a factor. People could argue that the emphasis on making a strong distinction between humans and other life forms is reinforcing anthropocentrism, the belief that humans are exceptional or supreme compared to other life forms. This anthropocentrism is argued to be the cause of human indifference to the ecology of the planet.[21] Even if humans are "hard wired" to see humans as distinct from other life forms, as described earlier, people arguing against anthropocentrism would argue that this is one of the innate human urges, like violence, that we should try to overcome. The barrier would reinforce this errant value and should be taken down. Of course, the primary threat to this barrier is that nobody would bother to build it in the first place, given that it is at or near the dystopian bottom of the slope.

The Liberal Eugenics Barrier

For the past two chapters I repeatedly pointed to the values of autonomy and beneficence as the reasons barriers could fall. Additionally, the debate has always included the value of nonmaleficence (avoiding harm)—no one advocates conducting an act of HGE unless it is safe.

The last barrier I will consider is the one that would be constructed at the limit of autonomy and non-maleficence, after all the upslope barriers have fallen. This is the perspective of mainstream bioethics, and bioethicists might not seem like they believe in barriers, but they do. Their perspectives often put *conditions* on HGE, such as that it must first be safe, but their barriers often remain obscure because they are so far downslope of where the debate is at present.

What I will call the liberal eugenics barrier holds that prospective parents have the autonomy to engage in any HGE they desire, short of harming others. The barrier is therefore supported by only two values: autonomy and non-maleficence. In contrast to traditional eugenics, which was coercive, this is voluntary eugenics, where parents autonomously decide what traits are diseases and what traits would be an "improvement" if changed. Beneficence is not really a supporting value since it is defined through the autonomous viewpoint of the parents.

Therefore, upslope of the barrier would be any autonomously chosen selection or modification that does not harm anyone else. Downslope would be coercion or a selection/modification that harms someone else. So, for example, were it possible to use HGE to modify your child's intelligence, this would be permissible, as long as the parents freely made the choice and in doing so did not harm others.

This barrier represents what is roughly taken to be the norms for a liberal democratic society: People have the autonomy to do whatever they wish as long as they do not harm others in the process. This concept is built into American law and is the basis of reproductive rights discourse in the United States. I have argued elsewhere that it is the basis of principlism, which is the underlying logic of the bioethics profession that now dominates these debates.[22]

Many participants in the debate have long advocated for this barrier. A typical advocate is philosopher John Harris—an invited speaker at the NASEM summit—who writes that "the burden of proof is not on those who would exercise this liberty or right to enhancement to show what good it does, rather, it is on those who would limit it to show how and to what extent its denial is necessary to protect either the exercise of a like liberty for all or is required to protect others or society from real and present harms or dangers."[23] ("Liberty" is roughly equivalent to "autonomy," at least for the purposes of this book.)

Similarly, lawyer John Robertson quotes John Stuart Mill as writing "the only purpose for which power can be rightfully exercised over any member of a civilized community, against his will, is to prevent harm to others." As Robertson summarizes it, procreative liberty is not absolute, but rather "there is a strong presumption in its favor, with the burden on opponents to show that there is a good case for limiting it," primarily that it harms others.[24]

So, in sum, above this barrier are almost all instances of somatic and germline genetic modification. Some participants in this debate might say that they thought this position *was* the bottom of the slope. But, there actually is terrain further down, which is where people can do whatever they want if they have power and wealth, regardless of the impact on those without power. This is a real threat to this barrier because it is analogous to how wealthy people often act in contemporary society on non-genetic issues.

Weaknesses of the Liberal Eugenics Barrier

The liberal eugenics barrier would have no continuity vagueness, but how the values of autonomy and non-maleficence would define the acts on both sides of the barrier would suffer from extreme similarity vagueness, potentially bringing down the barrier. Let us start with the vagueness of autonomy. If coercion is the opposite of autonomy, we can agree that having the state conduct a medical procedure on you without your knowledge or against your will, as happened in the pre–World War II eugenics movement, is coercion.[25] At the other endpoint, such a decision would be fully autonomous if the norms of society did not have a preference for what you did, the procedure was free to you, and your decision would have no negative impact on you or your child. The enormous area in-between these endpoints is vague. Certainly a decision that produces a financial burden is not truly autonomous, and social norms can be quite coercive. Would a society where the enhanced get special opportunities result in subtle coercion of the parents?

Consider the act of enhancing the intelligence of your child through selection or modification. Which side of the barrier is that on? On the one hand, people may be technically free to have children "the old-fashioned way." On the other hand, doing so may mean that their child's life chances are seriously reduced, imperiling their economic future. Was that a free decision? The ambiguity of which acts of HGE are really autonomously decided upon makes this barrier very susceptible to similarity vagueness.

The application of non-maleficence (harm to others) to acts surrounding the barrier is even more vague. There are two types of "harm": individual and social. Individual harms can be somewhat less vague, and there is high consensus that a physical harm to a person's body (like a broken leg) is an actual harm. Yet, even here, debate continues on whether parents can harm their offspring genetically by, for example, creating a child with a condition that everyone else thinks is a disability.[26] If this is what "harm" means, a barrier that is built far, far downslope, and which only excludes acts of HGE that result in the physical harm to others, could possibly hold. This is roughly what is easiest to demonstrate in contemporary US society outside the genetics debate—that your autonomous actions harmed my body.

Psychological harms are much more difficult to demonstrate because of subjectivity. For example, some argue that genetically designing children is harmful because it stops them from having an open future where they would fashion themselves to be the sort of person they want to be.[27] A child engineered to be a master musician would not really have an open future to pursue other things. If "harm" includes these notions, harm would be extremely susceptible to similarity vagueness.

What is worse for this barrier is that, as the NASEM report points out, the most important harms are of the second type—they are social and cultural and are thus "necessarily more diffuse."[28] The NASEM report discusses a number of these, such as a loss of a natural genome, a loss of human dignity, the rise of negative views of the genetically different, that humans will come to be thought of as more object-like or that we will come to live in a eugenic society.[29] Not only are these notoriously difficult to demonstrate ahead of time, there is no one individual to point to who has been harmed. In sum, this barrier probably suffers from terminal similarity vagueness.

Values That Would Support the Liberal Eugenics Barrier

The only value that would threaten the liberal eugenics barrier would be something like "might makes right," and that therefore the powerful deserve to be able to harm the unpowerful. People do not usually explicitly advocate this value. The primary threat to this barrier is not from values that would knock it over but from extreme vulnerability to slippery slope processes.

Despite its instability, it would be supported by a number of values that are very powerful in American culture. Indeed, it could be argued that this barrier *is* American culture, where people can do anything they want unless someone else can argue that it harms them. For example, in the United States you can listen to music in your home at eardrum damaging levels—unless it is bothering the people next door.

The dominance of these values in the HGE debate is well documented. A close reading of the arguments people make shows that there have long been people who have made intrinsic arguments against HGE, such as that to do so will make humans more like objects. But, after the bioethics profession came to dominate these debates, such people learned that for these intrinsic arguments to be taken seriously they must be translated into the language of harms. That is, one cannot argue that HGE is wrong because it makes all humans seem more like objects, one must argue that it harms the interests of people who do not want to live in a world where humans are more object-like.³⁰ Or, similarly, it could be intrinsically wrong to create a modified human that has great athletic ability, but for that argument to hold, it must be translated into either the individual-level harm of restricting the open future of the child or the collective harm of damaging competitive sports. It is clear that non-maleficence is dominant in these debates.

Conclusion

In this chapter I examined hypothetical barriers downslope of the median trait barrier discussed in the last chapter. The intuitively attractive and much discussed goals of medicine barrier would suffer from similarity vagueness so extreme that a decision about where to place it would be impossible. For example, one of the goals of medicine—at least in that it is taught in medical schools and engaged in by doctors—is cosmetic enhancement. The goals of medicine are potentially anything that has to do with the human body.

The family genes barrier is much sturdier and suffers from only mild similarity vagueness. It is, however, very far down the slope and would allow couples to select or install any trait that they could have produced through sexual reproduction. Moreover, the social impact of this barrier depends on how heritable traits actually are. If any couple could, one out of a 1,000 times, produce an Einstein, then the effects on social stratification will be large as many people would want an Einstein. If the possible traits the parents could produce in a child cluster more closely to the traits of the parents, then the impact on social advantage would be less.

The limits of humanity barrier is also quite sturdy and not susceptible to slippery slope processes. While a change in values could challenge the barrier, the main challenge is that this is so far downslope that few people would bother trying to create intellectual justification for it.

Finally, I considered the liberal eugenics barrier, which has long had many advocates in the debate, particularly in the bioethics profession. With this barrier, any modification or selection is acceptable as long as doing so does not harm another. This barrier is also very unstable, and, like the goals of medicine barrier, it could probably not be placed on the slope.

5
Conclusion

Will we as a society allow people to install genetic traits in their children that give them social advantage or "improve" them in general? Will we get to the point where most children from the upper socioeconomic classes are designed to have the "best" genes? This obviously depends upon technological developments, and it is possible that our abilities will never move beyond where they are now. It also depends upon politics, social movements, and corporations. However, another determinant of what we eventually allow will be the public bioethical debate, which largely consists of academics. Public bioethical debate does not have direct power over what happens but frames the options for others. It is therefore important to understand how public bioethical debate operates.

Participants in this debate can in theory say anything, but they have incentives to adhere to the established norms in the debate. One norm is that debates about technology and the human body are framed in terms of limits—"up until this act is ethical, beyond that act is unethical." Another is that these arguments have a "moral other" (like Nazis, Gattaca, or the Brave New World) with which to contrast your position. When combined with the discursive logic used by academics, such as rational consistency, these norms combine to produce a debate organized like a slope with the "moral other" at the bottom and barriers on it that stop us from ever reaching the bottom. Through close examination of how this slope works for the HGE debate, I hope that this book produces greater understanding for its participants and thus an improved debate.

Historical and Future Barriers on the HGE Slope

One of the goals of this book is to describe the past and cautiously predict the future of the HGE debate. These predictions follow the theory of slippery slopes and barriers in public bioethical debates which was developed partly through examining the history of the HGE debate. To recap this history, the earliest debate in the 1950s was that of the eugenicists, and it was not organized like a slope. By the early 1970s the slope began to form, with somatic gene therapy at the top and germline enhancement at the bottom. Two barriers were also created. One barrier was the disease/enhancement barrier, which separated the then obvious cases of monogenic inherited diseases like sickle cell from the eugenicists' dream of improving the intelligence of the human species. The other was the somatic/germline barrier that separated modifying the genes within the body of an existing person and modifying the genes of the human species.

The disease/enhancement barrier was strong because the acts on the two sides were utterly distinct. The only known and understood genetic diseases were those that were the result of a single gene gone wrong and which led to obvious disability in children. Downslope were enhancements that were the result of many interacting genes that were utterly not understood. With beneficence as an important value in the debate, upslope was the clear and obvious suffering of people with destructive genetic diseases, and below would be the desire of parents wanting to make their child the smartest in the class. Upslope was associated with medicine, downslope with the Nazis. This was a powerful barrier until increasing scientific knowledge made possible the acts that were immediately on either side, resulting in severe similarity vagueness.

The somatic/germline barrier was even stronger as the terrain it divided was who would be modified—the existing human body or the species. This barrier was also supported by influential values.

Non-maleficence supported the barrier because with somatic the effect of any mistake would end with the modified person's death. With germline (downslope) a mistake would go on to uncontrollably impact the human species. Moreover, above the barrier was following nature or God's will, and below was trying to control nature and/or God's plan. These two barriers became institutionalized and were strong for a few decades, making it easy for somatic gene therapy to proceed without fears that these applications would somehow slide into the dystopian bottom of the slope.

The barriers were not to last forever. A new profession entered the HGE debate named bioethics, and with the new profession came emphasis on the importance of a few values: autonomy, beneficence, non-maleficence, and justice. The theologians slowly became fewer in number, and the value of following nature or God's will was in decline. The somatic/germline barrier was slowly weakened as people in the debate argued that the value of beneficence justified not only acts above but also below the barrier. Similarly, others argued that the value of autonomy made the germline barrier obsolete because making children was reproduction, and parents should be able to determine the genetic qualities their children will have.

Both of these arguments represent the more difficult way to fell a barrier because they radically redefine a huge part of the slope. The radicalness of these claims is one reason why the germline barrier was still standing at the beginning of the 21st century. Also contributing was the fact that few people were actually pushing for germline HGE. The invention of PGD meant that almost anyone who was at risk of passing on a genetic disease to their children could use PGD instead of the unknown and thus risky HGE. Moreover, scientists could not even get somatic gene therapy to work, so the idea of gene editing embryos remained fanciful.

The disease/enhancement barrier was weakened by increased scientific knowledge and new innovations. When the barrier was made, the only traits that were understood were so obviously diseases that they were far, far above the barrier. "Enhancements" were so hypothetical that they remained the eugenicists' fantasies of improving traits like intelligence and empathy of the human species. But, by the 1980s, scientific innovation made it plausible to imagine intervening in traits that were close to the barrier. Scientists were beginning to understand that some people did not "have" genetic disease but had increased likelihood of developing diseases not previously thought of as genetic (like heart disease). Genes were also discovered that led to adult-onset diseases, further blurring the disease concept.

Moreover, the first somatic trials of gene therapy were actually to genetically enhance the cells in a person's body to fight cancer—which hopelessly muddled any clear distinction between disease and enhancement. Scholars in the debate did make various attempts to derive a value that would justify a demarcation between these different "disease" applications. Those attempts failed. That all said, these barriers survived to the end of the century because, despite decades of hope, scientists were still unable to get basic somatic gene therapy to work for single-gene traits that were universally considered to be serious genetic diseases. Without success in these consensual, somewhat understood, and severe diseases, modifying the "diseases" in the gray zone near the barrier remained too dangerous to pursue.

The invention of CRISPR technology at the start of the 21st century made interventions seem possible, and therefore made participants reconsider the viability of the long-standing barriers. The values in the debate had continued to change so that now there were few if any who would defend the idea of following nature or God's will, and instead the values were autonomy, beneficence, non-maleficence, and justice. This was evident by the main justification for maintaining the somatic/germline barrier, which was that doing so supported non-maleficence because germline was risky. This is very weak support because increasing scientific knowledge decreases risk and continually justifies moving beyond the barrier.

I closely examined the report of the NASEM committee that crystalized all the arguments against the barriers from previous decades and advocated pushing over the somatic/germline barrier. Previous decades had made it clear that the somatic/germline barrier was flawed because it was not only "the species" that existed downslope but "a baby" as well. Eugenicists of the 1950s were not focused on individual babies, only the species, but since then the field of reproductive medicine had developed that was only concerned with individual babies not the species. Species modification was an unavoidable and unintentional complication for those concerned with individual babies.

Technological developments meant that just upslope of the germline barrier was using PGD to avoid a trait like sickle cell anemia and just below was using HGE to modify a reproductive cell to avoid sickle cell anemia. Both technologies would impact the child and the species to the same extent, so for the barrier to hold, there would need to be a value that distinguished between "selecting" offspring (PGD) and "modifying" offspring through gene editing. A barrier *could* have remained between those two, were there someone in the debate who would support the value of following nature/God's will (which would support selecting but not modifying). Lacking such persons, there was extreme slipperiness between the couples who could use PGD to avoid sickle cell and a similarly motivated couple who could not use PGD and thus would need HGE. This was enough to bring down the germline barrier. The NASEM wanted to keep the barrier located at "serious disease," while acknowledging how vulnerable that barrier was to slippery slope processes.

That was the end of my history, and it is that history I used to identify the operation of the slope and barriers in the debate. I then used what we learned about the previous barriers to both create hypothetical barriers further downslope that appear less vulnerable to slippery slope processes and to evaluate downslope barriers that have been identified by scholars in the debate.

Given that the barrier proposed by the NASEM cannot hold after monogenic traits that are consensually considered to be diseases are successfully addressed with HGE, I created a strong barrier from the intellectual material in the NASEM report. While our abilities remain limited to single-gene traits where only rare variants are bad for you, the debate can rely upon the "most prevalent variant" barrier, where the most prevalent version of any gene can be selected for or edited in. When technology evolves to allow people to select or modify polygenic traits, the median trait barrier is necessary. In the median trait barrier, parents will be allowed to genetically modify their children (and by extension those children's descendants), but only by selecting or modifying reproductive cells so that the resulting baby has the median score on a polygenic trait. This is resistant to similarity or continuity vagueness because there is only one median score.

The new barrier is supported by the value of "justice" or "fairness." Below the barrier are those applications that would give a child social advantage, and above the barrier are applications that would level the genetic playing field. Again, this is because to gain social advantage you would have to be made above the median on some trait, which is not allowed with this barrier. Becoming average does not provide social advantage.

While this barrier has a fairly solid design, it is still vulnerable to the application of values that justify upslope acts to the acts below the barrier. Beneficence is clearly a threat because if scientists discover a non-prevalent variant that protects against disease, they would not be able to install it in someone's child. Indeed, the recent creation of Chinese twin girls who were supposedly given a rare variant that makes them resistant to disease already violates this barrier. Autonomy continues to be a threat as many debate participants continue to argue that anything that has to do with creating children should be only subject to the autonomous decisions of the prospective parents.

While I think that the median trait barrier is the strongest that can be built on the upper part of the slope, I examine proposed and hypothetical barriers further downslope. One very appealing barrier has been based in claiming that anything that fulfills the goals of medicine would be upslope, and those acts that do not are downslope. I conclude that this barrier would be susceptible to extreme slippery slope processes. To take but one example, the medical profession is engaged in cosmetic bodily enhancement, so if that is a goal of medicine, this barrier would be somewhere near the bottom of the slope, because germline editing for bodily cosmetics would be analogous to surgery.

In a second future scenario, biotechnology has improved to the point where many female eggs can be produced using pluripotent stem cells from the woman's skin. When fertilized, this would produce a large number of embryos that could have their genome sequenced. This would result in a type of PGD where a couple could identify the best embryo along some scale and gestate that one. If this PGD on a hundred embryos were to encourage the dismantling of the median trait barrier, and become common practice, the barrier for PGD would be at what is possible—the parents could use any combination of gene variants that they could themselves produce in a high number of tries. To place germline modification at the same point of the slope would require the family genes barrier, which would allow the modification to any set of variants that the parents could have produced on their own. While the barrier itself is only moderately susceptible to slippery slope processes, and being fairly supported by a set of remaining values, the barrier itself would be very far down the slope.

If the family genes barrier were to not hold, further downslope is the boundary of humanity barrier. In some future decade it may be possible to identify gene variants in other species that were useful to humans or variants that were just designed and synthesized.

A barrier not subject to continuity or similarity vagueness could be built at "the variant has existed in a human."

Such a barrier would be supported by the value of following nature/God, and nature/God has not yet resulted in any consequential gene sharing between humans and other life forms. It would also be supported by the fact that nearly all cultures distinguish between humans and other life forms so strongly that scientists have suspected that drawing these distinctions is part of our evolved brain. As usual, this barrier would be threatened by beneficence if there was some non-human animal gene that was beneficial to humans, and by autonomy, which would justify these human/animal chimeras on the grounds of reproductive freedom. I think that this barrier may hold, but there is very little terrain below it.

The final barrier is the liberal eugenics barrier, and it is the logical extension of the values used by the bioethics profession. With this barrier parents could edit their children's genes in any way they wanted, as long as they did not harm anyone else in the process. Many participants in the debate would assume this *is* the bottom of the slope, but there is a small bit of terrain below where people with power modify their offspring regardless of the effects on others. This barrier is highly susceptible to slippery slope processes because autonomy and harm remain elusive concepts. Despite its instability, this barrier is supported by influential values such as political liberalism. In fact, this barrier is located at the bare minimum in liberal democratic societies, which is itself built into the US legal system, where you are generally allowed to do whatever you want as long as you are not impinging upon others.

Can Any Barriers Hold?

While there was once consensus about the somatic/germline and disease/enhancement barriers, the debate is now without any agreed upon limits. I suggest a number of possible barriers.

However, at each rewrite of this book, each barrier I described from the germline barrier down the slope looked less and less stable. Even barriers that were not susceptible to slippery slope processes seemed particularly threatened by powerful values in this debate, such as beneficence. In the presentations I have made on this project, and in the informal review of the manuscript by colleagues, no one identified strengths I had ignored—only more reasons why a barrier could not hold. Moreover, I have not even mentioned an additional pressure on any barrier, which is that even if the United States creates barriers on the slope, another country will not, and soon enough people will challenge why similarly situated people who only differ in country of residence have differential access to HGE. Therefore, while I originally thought I was carefully describing how people in the debate could defend a barrier, I may be carefully describing how each barrier will eventually fall in turn. Time will tell.

Some scholars will agree with the latter interpretation and have already written that we should give up on creating what I call barriers on the slope, perhaps not because this is good, but because no barrier will hold. I take that to be the position of Eric Juengst, who concludes that "the emergence of enhancement applications of gene editing is inevitable," so "instead of trying to come to consensus on incommensurable community worldviews and personal values, perhaps it would be better to encourage the public to prepare for a world in which gene-edited enhancements and occasional inadvertent germ-line changes are a reality and to discuss the human rights protections that the variegated inhabitants of such a world will need."[1]

We would then stop "policing science," he continues, and create a society where the onset of HGE will not result in Gattaca. This would "turn the conversation toward more mundane matters: combating genetic discrimination against both the enhanced and the unenhanced, leveling playing fields by changing the rules of our games rather than by excluding those who want to play, and living up to our commitment to treating people as moral equals in spite of their biological diversity."[2]

I would say that, before we give up on barriers—on limits—we have to at least acknowledge that the task that Juengst calls for is far, far more difficult than is "policing science." Implicitly, Juengst's dystopia is the liberal version of severe inequality, but creating a world where people are not able to use the genetics of their children to solidify advantage requires, essentially, eradicating structural inequality. We would need to make the United States more like Denmark where the family you were born into has less of an effect on your life outcome—a noble goal, but certainly a much more difficult task then passing a law restricting how people are allowed to modify their children. As daunting as defending barriers appears to be, it may be the easier way to forward your values.

Could Solid Barriers Be Anchored by Authority?

I am sure you have noticed that the barrier and slope metaphor is very concerned with rules and does not permit discretion or any process where people could just come to a rough substantive compromise on, for example, what a disease is. This is because this is focused on US political culture. In Europe, it is possible that barriers could survive, despite being vague with a slippery slope around them, because those societies tend to trust expertise and judgment. As a simple example, studies that focus on this difference between the United States and Europe would suggest that in Europe people would be more willing to say that doctors can decide what a disease is.[3]

But, that is not how American culture works. It has been noted that "the United States relies on rules to control the exercise of official judgement to a greater extent than any other industrialized democracy," and therefore "there is more need for objectivity than in Britain or France." Moreover, in European countries, experts "are trusted to exercise judgment wisely and fairly. In the United States, they are expected to follow rules."[4] It is part of the American DNA, so to speak, to not trust the government. In particular, it is hard to imagine Americans saying they would trust government officials to make discretionary moral judgments. Barriers that are not susceptible to slippery slope processes may not be agreed to by the public, but they suggest that any official who adopts them for use by their agency is not using their own values to make judgments.

A good example of government policymakers being able to use their own judgment and discretion in defining disease is in the United Kingdom, where the government has determined what a "serious" genetic condition is, without actually precisely defining serious. A couple can only obtain PGD if the condition they are trying to avoid is on the list of the regulator, named the Human Fertilisation and Embryology Authority. At present there are about 600 genetic conditions on their list. When they decide to allow a new trait to be avoided with PGD they examine factors such as "how serious the condition is" and "the testimony of people affected by the condition."[5] Critics will note that the list includes "diseases" like dwarfism and deafness.

A criterion such as "serious" is hopelessly susceptible to slippery slope processes, but a society that trusts in experts and elites can rely upon their opinion about seriousness to create a barrier that is at least more resistant to slippage. Therefore, this analysis is more relevant to the United States than it is to other countries.

Rejecting the Direction of the Slope

I have only been talking about the primary structure in the HGE debate, which is a slippery slope down to Gattaca and the Brave New World. It is important to remind ourselves that there are people in the debate who have a different view of what would happen if technology continues to develop and more and more applications are allowed. They do not see a dystopia at the end, but a utopia. For these advocates, the slope does not go down but up. Every technological development and every rejection of a constraint on a couple's freedom is another step up toward the utopian top of the slope. Such an advocate could write a book about the construction of steps up the slope, not barriers stopping a slide down.

In the earliest years of the debate, the eugenicists would have argued that the entire point of the debate was to climb the slope up to the utopia where humans had transformed themselves into a superior species. The eugenicists still exist in this debate, but now they call themselves "transhumanists," where the point is to rise above the limitations of this version of the human and achieve humanity 2.0—an improved version of the species. They tend to downplay the problems the critics identify and focus on how such a society would be superior.

Consider transhumanist Nick Bostrom's famous "Letter from Utopia," which is an argument in a form of a letter from future humans in the genetically perfect society to the current limited humans, trying to convince the present humans to perfect themselves. "Your body is a deathtrap" the writer implores, "you are lucky to get seven decades of mobility," which is "not sufficient to get started in a serious way, much less to complete the journey." Indeed, immortality is the goal because "any death prior to the heat death of the universe is premature if your life is good." The writer implores the present day humans to "upgrade their cognition" because "your brain must grow beyond any genius of humankind, in its special faculties as well as its general intelligence, so that you may better learn, remember, and understand, and so that you may apprehend your own beatitude." Pleasure is universal in the transhumanist utopia after the species has been improved. "A few grains of this magic ingredient are worth more than a king's treasure, and we have it aplenty here in Utopia. It pervades into everything we do and everything we experience. We sprinkle it in our tea."[6] The transhumanists remind us that while I am describing the mainstream public bioethical debate about HGE, there are others who see the debate from a totally opposite perspective.

General Conclusions About Slippery Slopes from the HGE Case

The case of HGE may offer insight into other debates. Like many debates in liberal democratic societies, public bioethical debates are set up as demarcating the limits of individual freedom, so typically they are structured as "up to this point you are free to do as you wish, below this point is unethical and you are not free to act." Consistent with the depiction of the vertical dimension in Western culture, up is good (e.g., heaven above) and down is bad (e.g., hell below). The continuum from good to bad is therefore tilted with "up" representing the most meritorious act, and "down" representing the most reprehensible act, creating a slope.

Debaters metaphorically step onto the top of the slope by advocating the most meritorious act, but want to make sure they do not slip down the slope to the unethical acts below—and certainly not to the evil "other" rhetorically represented by the bottom of the slope. The danger is that the slope is slippery because of the absence of a barrier between the meritorious step and the one below it. There are two reasons that a barrier does not exist. The first is "continuity vagueness," when there can be no argument in the difference between, say, 17 and 18. The second is "similarity vagueness," where distinction is not possible because of the only distinction that can be made between the steps does not hold ethical weight (e.g., a disease of the liver vs. a disease of the kidneys). Slipping is when the consensus of what is ethical in the debate moves to the next step down the slope.

The terrain of the slope will differ for each debate. In the abortion debate the terrain has traditionally been fetal age and the motivation of the woman to have an abortion. In the HGE debate the terrain at least initially was genetic trait and who would be modified (the target). In the debate about aid in dying, the top has been something like physician-assisted suicide for an elderly, mentally competent patient who faces immanent death from an incurable disease that causes great suffering. At the bottom would be euthanasia of people society decides are not of value. Public bioethical debate is particularly susceptible to slippery slopes because the participants put a very heavy weight on logical consistency.

If two steps on the slope are too similar to distinguish, we must ask, similar by what standard? Values justify actions at each location on the slope, so for the debate to slip one step down the slope requires that the two actions be the same according to the same value. The values also define the terrain or what should be noticed about the terrain. For example, if the only value was beneficence, the genetic traits would be ordered by the suffering they cause. If the only value was following nature, the genetic traits would be ordered by how "natural" they are.

Slippage on the slope can be stopped with the creation of a barrier that separates two applications of HGE that cannot be made the same. I identified a number of ways that a barrier can be felled, with some being more difficult than others. A nearly insurmountable barrier is defined so that the applications on the two sides of the barrier are not subject to continuity or similarity vagueness *because* the values that cover the debate provide inverse evaluations of the acts on the terrain upslope and downslope. This is represented by frame 1 in Figure 1.1.

The most difficult way to fell the barrier is to claim that a value justifies not only the acts upslope of the barrier, and those just below the barrier, but also the acts far below the barrier. An example is the value of autonomy, which in the early 1990s was argued to apply to acts far below the somatic/germline barrier, to the use of germline for any desired trait. To argue against a barrier like this requires explicitly or implicitly claiming that the values that had long held up the barrier are wrong—a more revolutionary stance that is less likely to be successful. In the history of the HGE debate, this method of felling a barrier always threatens, but has never actually happened. Frame 2 in Figure 1.1 depicts a barrier facing this threat.

A weaker barrier occurs when the terrain just below the barrier is given a positive evaluation by the value that positively evaluates the acts above the barrier. This is easier to accomplish because it does not require declaring that the previous values in the debate are wrong—only that a value that everyone otherwise agrees with has slightly more expansive applications than previously recognized. Since values define the terrain, this slightly redefines the terrain right below the barrier, making it similar to that above, creating either similarity or continuity vagueness. The vagueness results in the debate slipping over the top of the barrier and justifying acts immediately below. The barrier is then gone. Barriers that were already very weak due to continuity or similarity vagueness (e.g., disease/enhancement) will quickly fall. Those otherwise not susceptible will require more effort (e.g., the somatic/germline barrier). (Frame 3 of Figure 1.1.)

Slopes Where Downslope Actions Are Not Yet Possible

I have not conducted a comparative analysis of slippery slopes in another debate, but I can speculate about possible differences. I suspect that upon systematic comparison we would find a difference between debates where every act along the slope is currently possible and those where the downslope acts would require further scientific innovation or knowledge. With abortion and euthanasia, every location on the slope is currently possible, whereas with HGE the lower locations require scientific developments to come to fruition.

A slope that requires innovation to get lower on the slope will always have a "scientific reality" barrier that comes in strong and weak forms. The strong form is the argument that we do not need to worry about the acts far downslope because they will *always* be impossible or are so many decades in the future that it is not worth thinking about now. An example is arguing that the human traits that people are interested in enhancing like intelligence, empathy, and so on will always be mysterious with no clear genetic causation. Another example is arguing that even if we understood the 100 interacting genes that lead to what is called "intelligence," it will always be scientifically impossible to edit 100 genes at once. The problem with the strong form of the scientific reality barrier is that it crumbles with our advancing knowledge and is not anchored in any values. Put differently, the barrier has no advocates, it is just based in "facts."

In the HGE debate, the weak form of the scientific reality barrier is often described in terms of safety. It is almost a mantra in these debates over the past 50 years for participants to say that a particular application of HGE should not be implemented until it is safe to do so. In fact, all of the post-CRISPR statements I reviewed at the beginning of Chapter 3 generally say that we should not go beyond the location of the germline barrier on the slope because it is not safe. But, the danger for the debate is that this weak scientific reality barrier is often the only barrier placed on the slope—particularly by scientists, who often see safety as the primary or even the only moral issue.

Like the strong version, this weak version of the scientific reality barrier tends to fall with regularity because human knowledge is advancing at such a rapid clip. The NASEM report deserves particular credit for not relying upon safety and for substantively addressing the question of what should happen when our knowledge increases.

Defenders of the use of scientific reality barriers will point to the fact that we cannot have precise ethical debates about technologies that do not yet exist. True enough. Ideally, when deciding where to construct a barrier on the slope people should weigh the probability of the scientific reality barrier crumbling. Of course, I am not the first one to point out that if we treat the scientific reality barriers as barriers beyond which we cannot look, then when they fall we will not have prepared ourselves for the terrain below. At worst, the scientific reality barriers take on an ideological purpose of obscuring the scary bottom of the slope to increase the likelihood of moving past barriers further upslope.

I suspect that slopes where the bottom is not yet possible generate stronger reactions because the nature of the bottom is actually unknown. Will HGE be able to influence the morality of one's children? Will HGE be used to create a class of super-smart humans? People can read their fears into the unknown, which offers more reasons to avoid reaching the bottom. Contrast this to the case of euthanasia, where it is fairly easy to see what would happen at the bottom of the slope with the involuntary euthanasia of "socially useless" individuals because this actually happened in the middle of the last century in Nazi Germany.

Objectivity and Strong Barriers

The received wisdom in the field of science studies is that what is considered a "fact" is that which seems to be the most distant from human interpretation. It is notable that the barriers that I argue are the least susceptible to continuity or similarity vagueness slippery slope mechanisms are those built on what are considered to be objective facts about nature and not on human generated concepts. For example, the somatic/germline barrier was based on the scientific fact that somatic human cells do not move everywhere and blend with the reproductive cells of the body. The cells in a human's liver, bone marrow, or brain do not merge with their sperm or egg cells, and therefore the distinction between somatic HGE and germline HGE is based in biological reality. (That such a divide is morally relevant is obviously a human construction.)

Similarly, it is a scientific fact that humans are *extremely* similar genetically, with a few variants that make their bodies distinct from others. It is then objectively true that there are some gene variants that are more common than others, and therefore traits will tend to exist in bell-shaped distributions. Even if how we value difference is subjective, it does appear that there are biological outliers to any trait we can come up with. Therefore, a barrier based on the fact that there is going to be a median human on a trait will be fairly strong.

Barriers that are not based in biological reality are weaker, presumably because they are based on human ideologies that change rapidly, differ by subculture, and are more clearly based in the extremely imprecise medium of language. "Disease" and "harm" are the best two examples. One problem with these objective, biology-based barriers is that such objective distinctions are probably few and far between, and therefore can only be placed in a few locations on a slope. Indeed, in the last chapter I tried to come up with as many downslope barriers as I could based on biological distinctions and found few.

What Is to Be Done?

For the Participants in the Public Bioethical Debate

I would like participants in the public bioethical debate about HGE to reconsider how they make their arguments by explicitly considering the mechanisms identified in this book. Without this change, arguments at best mislead the reader by implying that the position taken is viable. At worst they give the reader a false sense of security that they can walk onto the slope without sliding.

An ideal process would start with, first, deciding your true barrier on the slope, which is inevitably based on your values. This is how participants in the debate currently function—the point is to come up with the most morally defensible barrier and then try to convince others of it. My recommended second step deviates from standard practice—acknowledge your true barrier and then analyze whether that barrier can hold.

To take the extreme case, let's say your barrier is based on allowing HGE for diseases but not enhancements. You *have* a definition of disease because you have particular values, and thus you have created a legitimate barrier—it is just a barrier that will not survive in the pluralism of values that exists in the broader HGE debate. The writer should then either state why their barrier based on "disease" is not susceptible to the mechanisms I identify in this book, or if it is susceptible, simply state that what is being offered is a barrier that is not expected to survive.

For example, I think many debate participants would like for the goals of medicine barrier to be defendable, because they think what are *current* medical practices and norms roughly coincide with their values. If what I say is correct, by the time we would get that far downslope with our abilities, that barrier would not even be strong enough to build in the first place, as the profession of medicine will be far beyond treating sickle cell anemia. Part of the argument of such a person would be to justify the goals of medicine barrier, and part would be to argue that it is defensible. Similarly, attention to my analysis would require acknowledgment that the strong and weak scientific reality barriers are not barriers at all, but rather just slide downslope as scientific knowledge increases.

Ideal practice in the debate would also start with describing one's barrier; acknowledging if it cannot hold; and if it cannot, then explicitly abandoning it to support the most defensible barrier upslope or downslope. If the practice of supporting a defensible barrier instead of one's ideal barrier were to take hold, this would take participation in public bioethical debate another step away from academic purity and toward politics. But, this has always been what separates public bioethical debate from fields like philosophy. <u>In philosophy the point would be to decide what should be, in public bioethical debate the point is to decide what should be given social and political constraints.</u> In public bioethical debate, the question to ask is: Why spend the time defending a position that cannot hold?

For the General Public

Perhaps most readers of this book will not be participants in the public bioethical debate. Ultimately, public bioethical debate is supposed to contribute to policymaking that is done in the public's name, and the public is implicitly consenting to such policies. For example, the National Institutes of Health, part of the executive branch and controlled by our elected president, has a policy of not supporting germline HGE. That policy is a (very) small part of what you are voting for when you vote for our national elected officials who appoint the head of the National Institutes of Health and other decision-makers.

I hope that the lesson for the general public is to examine the barriers being proposed for these policies and ask whether they are viable. If the government proposes a barrier that cannot hold, and you are opposed to the acts below that barrier, you should identify the barrier as non-viable and say so. It is important that citizens, like academics, find a defensible barrier and stick to it. This will become even more important as our technological abilities with HGE improve.

[margin note: But if none are ultimately viable as you suggest at times?]

Notes

Chapter 1

1. Marilynn Marchione, "Chinese Researcher Claims First Gene-Edited Babies," Associated Press, November 26, 2018.
2. Google dictionary.
3. For the definitive histories of this movement, see Daniel Kevles, *In the Name of Eugenics: Genetics and the Uses of Human Heredity* (Berkeley: University of California Press, 1985); Diane B. Paul, *Controlling Human Heredity: 1865 to the Present* (Amherst, NY: Humanity Books, 1995).
4. Robert L. Sinsheimer in Kevles, *In the Name of Eugenics*, 267–68.
5. John H. Evans, *Playing God? Human Genetic Engineering and the Rationalization of Public Bioethical Debate* (Chicago: University of Chicago Press, 2002).
6. For a laypersons' summary of the technology, see National Academies of Sciences, Engineering, and Medicine, *Human Genome Editing: Science, Ethics, and Governance* (Washington, DC: National Academies Press, 2017).
7. Hong Ma et al., "Correction of a Pathogenic Gene Mutation in Human Embryos," *Nature* 548 (2017): 413–19.
8. Michael Specter, "The Gene Hackers," *New Yorker*, November 11, 2015, https://www.newyorker.com/magazine/2015/11/16/the-gene-hackers.
9. Motoko Araki and Tetsuya Ishii, "International Regulatory Landscape and Integration of Corrective Genome Editing into in Vitro Fertilization," *Reproductive Biology and Endocrinology* 12, no. 108 (2014): 1–12.
10. A. Buchanan, D. W. Brock, N. Daniels, and D. Wikler, *From Chance to Choice: Genetics and Justice* (New York: Cambridge University Press, 2000), 23.
11. Quoted in: Anna Almendraia, "World's First Gene-Edited Babies Could Set Genetic Science Backward, Experts Worry," *Huffington Post*, November 27, 2018.
12. Antonio Regalado, "Rogue Chinese CRISPR Scientist Cited US Report as His Green Light," *Technology Review*, November 27, 2018.

13. Katie Hasson, "Senate HELP Committee Holds Hearing on Gene Editing," *Biopolitical Times*, November 15, 2017. https://www.geneticsandsociety.org/biopolitical-times/senate-help-committee-holds-hearing-gene-editing-technology.
14. Beyond examining and explaining key moments in the past and future debate, I aspire to produce a type of mid-level theory. The purpose of mid-level theory is to help the reader look at the same data they are familiar with in a new and hopefully enlightening way. For the sociological reader, the inspiration I have in mind is Andrew Abbott's 1988 book *The System of Professions*. Andrew Abbott, *The System of Professions: An Essay on the Division of Expert Labor* (Chicago: University of Chicago Press, 1988). Before reading his book, the scholars of the professions would have all been aware of the history of professions like medicine and nursing, but through theoretical generalization and the introduction of a few mid-level concepts like "subsidiary profession," all of these professions were seen in a new way. My aspiration is that even the readers who have participated in the HGE debate for decades will be able to see what they already know in a new light, and that this in turn will provide for further insight.
15. Evans, *Playing God?*
16. There is a large literature describing the different types of slippery slope arguments. Douglas Walton, *The Slippery Slope Argument* (New York: Oxford University Press, 1992); Douglas Walton, "The Basic Slippery Slope Argument," *Informal Logic* 35, no. 3 (2015): 273–311; Eugene Volokh, "The Mechanisms of the Slippery Slope," *Harvard Law Review* 116, no. 4 (2003): 1026–137; Mario J. Rizzo and Douglas Glen Whitman, "The Camel's Nose Is in the Tent: Rules, Theories, and Slippery Slopes," *UCLA Law Review* 51 (2003): 539–92; Wibren van der Burg, "The Slippery Slope Argument," *Ethics* 102, no. 1 (1991): 42–65; David Albert Jones, "Is There a Logical Slippery Slope from Voluntary to Nonvoluntary Euthanasia?" *Kennedy Institute of Ethics Journal* 21, no. 4 (2011): 379–404.
17. Penney Lewis, "The Empirical Slippery Slope from Voluntary to Non-Voluntary Euthanasia," *Journal of Law, Medicine and Ethics* 35, no. 1 (2007): 197–210; Barron H. Lerner and Arthur L. Caplan, "Euthanasia in Belgium and the Netherlands on a Slippery Slope?" *JAMA Internal Medicine* 175, no. 10 (2015): 1640–41.
18. Jones, "Is There a Logical Slippery Slope from Voluntary to Nonvoluntary Euthanasia?," 384.
19. The argument is that "as a result of psychological and social processes," we will be more likely to accept B. van der Burg, "The Slippery Slope

Argument." As Holtug writes, "it is essential to an empirical version of the slippery slope argument that the causal connection between the action proposed and the undesirable outcome at the end of the slope be specified. Also, we need a reason to believe that the causal chain is probably enough to make it reasonable for us to refrain from making the first move." Nils Holtug, "Human Gene Therapy: Down the Slippery Slope?" *Bioethics* 7, no. 5 (1993): 416.
20. Volokh, "The Mechanisms of the Slippery Slope," 1031.
21. Volokh citing philosopher Douglas Watson in "The Mechanisms of the Slippery Slope," 1077.
22. See Albert R. Jonsen, *The Birth of Bioethics* (New York: Oxford University Press, 1998); David J. Rothman, *Strangers by the Bedside: A History of How Law and Bioethics Transformed Medical Decision Making* (New York: Basic Books, 1991).
23. Veikko Launis, "Human Gene Therapy and the Slippery Slope Argument," *Medicine, Health Care and Philosophy* 5 (2002): 169–70; Douglas Walton, "The Slippery Slope Argument in the Ethical Debate on Genetic Engineering of Humans," *Science and Engineering Ethics* 23, no. 6 (2016): 1508; Holtug, "Human Gene Therapy," 402.
24. Jonathan Haidt, "The New Synthesis in Moral Psychology," *Science* 316 (2007): 998–1002.
25. Leon R Kass, "New Beginnings in Life," in *The New Genetics and the Future of Man*, ed. Michael Hamilton (Grand Rapids, MI: Eerdmans, 1972), 785.
26. Julian Savulescu, "Procreative Beneficence: Why We Should Select the Best Children," *Bioethics* 15, no. 5/6 (2001): 413–26.
27. Mario J. Rizzo and Douglas Glen Whitman, "Little Brother Is Watching You: New Paternalism on the Slippery Slopes," *Arizona Law Review* 51 (2009): 691.
28. Rizzo and Whitman, "Little Brother Is Watching You," 691.
29. Society has worked out a functional compromise on the driving age of 16 in most US states, which is susceptible to continuity vagueness. That is, why not 15, 11, or 9 or 17, 21, or 23 years of age? Such a functional compromise only works when there is a consensus about the acceptability of the endpoints, and the endpoints are within a very small band. That is, nobody thinks that a 9-year-old should be driving, and nobody thinks that the average 21-year-old is incapable of driving. This is not the case in the HGE debate.
30. Rizzo and Whitman, "Little Brother Is Watching You," 691.

31. Legal theorists have assumed, as one puts it, that "humans arguably have a tendency to psychologically assimilate closely related cases." See Eric Lode, "Slippery Slope Arguments and Legal Reasoning," *California Law Review* 87, no. 6 (1999): 1513. Indeed, psychologists have produced evidence that these small steps are what drive the cognitive versions of these slippery slopes. Defining a slippery slope as a situation where "a presently unacceptable proposal ... will (by any number of psychological processes ...) in the future be re-evaluated as acceptable," psychologists call this a cognitive "category boundary re-appraisal mechanism." Imagine a cognitive category of "severe disease." There is a secular process of expansion, and these scholars note that it is fundamental to the contemporary consensus in psychology that "encountering instances of the category at the category boundary will extend that boundary for subsequent classifications. There is, then, a feedback loop inherent in the classification of new data into an existing category, whereby that classification also affects and alters the category itself." Their experiments demonstrate "how slippery slopes may rest on a category boundary extension process." They conclude: "the probability of predicted outcome following from an initial proposal appears to be directly related to the empirical similarity between the initial proposal and the predicted outcome." See Adam Corner, Ulrike Hahn, and Mike Oaksford, "The Psychological Mechanism of the Slippery Slope Argument," *Journal of Memory and Language* 64 (2011): 135, 136, 143. Similarity is the grease, and we have two qualities that make it easier to make similarities along the slope: continuity vagueness and similarity vagueness.
32. Rizzo and Whitman, "Little Brother Is Watching You," 692.
33. Casuistry, also known as "case-based reasoning," is based on analogies to previously adjudicated cases. With casuistry the analyst looks to already established cases, and if the new case is analogous to the established case, the ethical evaluation of the two cases is the same.
34. Albert R. Jonsen, "Casuistry as Methodology in Clinical Ethics," *Theoretical Medicine* 12 (1991): 297.
35. Tom L. Beauchamp and James F. Childress, *Principles of Biomedical Ethics*, 7th ed. (New York: Oxford University Press, 2013).
36. Rizzo and Whitman, "Little Brother Is Watching You," 692.
37. One of the components of slippery slope arguments that garners criticism is that they are not contingent enough and often imply inevitability. One reason is that they imply a closed system, where the only force changing the values used in the debate are the actions upslope. For example, the

arguments imply that a more expansive use of euthanasia is due *only* to the attitude-changing effect of less expansive uses of euthanasia upslope. But, when applied to a society, slippery slope arguments have to account for time, as well as the fact that people do not only think about euthanasia in their daily lives. Therefore, my argument accounts for the fact that values used in society change over time for reasons that may have little to do with upslope actions. For example, a more expansive use of euthanasia is partly the result of the growing strength of the value of personal autonomy—a growing strength that is only partially the result of what the public learned from the upslope use of euthanasia. This makes the slippery slope less deterministic, which makes it a more realistic method of analysis.

38. I have elsewhere established that the profession of bioethics claims it makes arguments on behalf of the public's values—as a stand-in for what the public would say if the public had time to get up to speed on the science and participate in these debates. This is understandably never made explicit, but it exists in theories such as bioethicists representing "the common morality," that demographically diverse commissions represent the public, or that by limiting perspectives to those that are purportedly "neutral" to the pluralistic groups in society. See John H. Evans, *The History and Future of Bioethics: A Sociological View* (New York: Oxford University Press, 2012), 43–48, 68–70.

39. Quote in Jones, "Is There a Logical Slippery Slope from Voluntary to Nonvoluntary Euthanasia?," 385. I suspect that at present Harris would not want to be associated with my argument.

40. In Sissela Bok, "The Leading Edge of the Wedge," *Hastings Center Report* 1, no. 3 (1971): 9.

41. Allen Buchanan et al., *From Chance to Choice: Genetics and Justice* (New York: Cambridge University Press, 2000), 13–14.

42. Rizzo and Whitman, "Little Brother Is Watching You," 738.

43. James F. Keenan, S.J., "Openness, with Caution and Suspicion, About Human Enhancement," in *Contemporary Controversies in Catholic Bioethics*, ed. J. T. Eberl (New York: Springer, 2017), 20. I think in the actual public sphere it is required that we be truthful about our reasons to support or oppose an action, so it is therefore acceptable to oppose some action that you are not opposed to in and of itself—as long as one states the reason is because it is the only defensible position near your actual position. I take this as morally akin to voting for a presidential candidate one does not fully agree with, but who is better than the alternatives.

Chapter 2

1. I covered this history from a macro-perspective in John H. Evans, *Playing God? Human Genetic Engineering and the Rationalization of Public Bioethical Debate* (Chicago: University of Chicago Press, 2002). This chapter re-analyzes this material for the micro slippery slope analysis.
2. Daniel Kevles, *In the Name of Eugenics: Genetics and the Uses of Human Heredity* (Berkeley: University of California Press, 1985), ix.
3. Kevles, *In the Name of Eugenics*, 34.
4. Kevles, *In the Name of Eugenics*, 96ff.
5. Kevles, *In the Name of Eugenics*, 97.
6. Kevles, *In the Name of Eugenics*, 110–11.
7. Melissa J. Wilde and Sabrina Danielsen, "Fewer and Better Children: Race, Class, Religion, and Birth Control Reform in America," *American Journal of Sociology* 119, no. 6 (2014): 1710–60; Christine Rosen, *Preaching Eugenics: Religious Leaders and the American Eugenics Movement* (New York: Oxford University Press, 2004).
8. Kevles, *In the Name of Eugenics*, 164.
9. Kevles, *In the Name of Eugenics*, 118.
10. Hermann Muller, "Genetic Progress by Voluntarily Conducted Germinal Choice," in *Man and His Future*, ed. Gordon Wolstenholme (London: J. & A. Churchill, 1963), 11.
11. Julian Huxley, "Eugenics in Evolutionary Perspective," *Perspectives in Biology and Medicine* (Winter 1963): 173.
12. Robert L. Sinsheimer in Kevles, *In the Name of Eugenics*, 267–68.
13. Tracy M. Sonneborn, *The Control of Human Heredity and Evolution* (New York: Macmillan, 1965), viii.
14. Paul Ramsey, *Fabricated Man: The Ethics of Genetic Control* (New Haven, CT: Yale University Press, 1970), 81.
15. Kelly Moore, "Organizing Integrity: American Science and the Creation of Public Interest Organizations, 1955–1975," *American Journal of Sociology* 101, no. 6 (May 1996): 1601.
16. Susan Wright, *Molecular Politics: Developing American and British Regulatory Policy for Genetic Engineering, 1972–1982* (Chicago: University of Chicago Press, 1994), 37.
17. Michael Hamilton, *The New Genetics and the Future of Man* (Grand Rapids, MI: William B. Eerdmans, 1972).
18. Bentley Glass, "Science: Endless Horizons or Golden Age?" *Science* 171 (1971): 28.

19. Bentley Glass, *Science and Ethical Values* (Chapel Hill: University of North Carolina Press, 1965), 59–60.
20. Bernard D. Davis, "Prospects for Genetic Intervention in Man," *Science* 170 (1970): 1279.
21. Davis, "Prospects for Genetic Intervention in Man," 1280.
22. Davis, "Prospects for Genetic Intervention in Man," 1279, 1280.
23. Theodore Friedmann and Richard Roblin, "Gene Therapy for Human Genetic Disease?" *Science* 175 (1972): 949.
24. H. Vasken Aposhian, "The Use of DNA for Gene Therapy—The Need, Experimental Approach, and Implications," *Perspectives in Biology and Medicine* 14 (1970): 106–7.
25. LeRoy Walters and Julie Gage Palmer, *The Ethics of Human Gene Therapy* (New York: Oxford University Press, 1997), xvii; Eric T. Juengst, "Can Enhancement Be Distinguished from Prevention in Genetic Medicine," *Journal of Medicine and Philosophy* 22 (1997): 125.
26. W. French Anderson, "Genetic Therapy," in *The New Genetics and the Future of Man*, ed. Michael Hamilton (Grand Rapids, MI: William B. Eerdmans, 1972), 109.
27. Paul Ramsey, "Genetic Therapy: A Theologian's Response," in *The New Genetics and the Future of Man*, ed. Michael Hamilton (Grand Rapids, MI: William B. Eerdmans, 1972), 157–75.
28. For my simple purposes, this means genetic traits where typically each of the parents is a "carrier" of the trait in that they have one of the alleles (versions of the gene) that leads to the disease, but their other allele of that gene is "normal." If someone has both copies of the disease version of the gene they manifest the disease. Therefore, in Mendelian inheritance, on average 25% of the offspring of two carriers will have two copies of the disease-causing allele and thus manifest the disease. Fifty percent of the offspring will be a carrier, like the parents were. Twenty-five percent will not be a carrier but will have two copies of the "normal" allele.
29. Davis, "Prospects for Genetic Intervention in Man," 1280.
30. W. French Anderson and John C. Fletcher, "Gene Therapy in Human Beings: When Is It Ethical to Begin?" *The New England Journal of Medicine* 303 (1980): 1296. Similarly, the chair of the Working Group on Human Gene Therapy of the Recombinant DNA Advisory Committee (RAC) of the National Institutes of Health wrote in 1986: "For the time being, moreover, only the simplest kinds of genetic defects are considered to be good candidates for gene therapy. These are single-gene defects which are recessively inherited and which result in the lack of an essential enzyme. The

production of such enzymes is almost a mechanical function, necessary to be sure, but shared by humans with many other members of the animal kingdom. Thus, any allegation that current gene therapy efforts represent 'tampering' with 'distinctively human characteristics' is simply fallacious." LeRoy Walters, "The Ethics of Human Gene Therapy," *Nature* 320 (1986): 225.

31. An example of this reasoning comes from prominent gene therapy pioneer W. French Anderson, who argued against somatic enhancement because the risk/benefit ratio was wrong: "In somatic cell gene therapy for an already existing disease the potential benefits could outweigh the risks. In enhancement engineering, however, the risks would be greater while the benefits would be considerably less clear." W. French Anderson, "Genetics and Human Malleability," *Hastings Center Report* 20 (1990): 23.

32. As the head of the Human Gene Therapy Subcommittee of the RAC wrote in 1986: "The chief technical questions surrounding gene therapy involve the comparison of potential benefits and harms, or, in contemporary language, risk assessment. The assessment process begins with an evaluation of the genetic disease to be treated. At this stage the central question is: what kinds of morbidity and what morality rates are associated with the disease? If the disease is severe for the individuals affected by it, one can proceed" with further consideration. Walters, "The Ethics of Human Gene Therapy," 225.

33. Kimmelman notes that by the time the trial actually began there was a more conventional treatment for this disease. Jonathan Kimmelman, *Gene Transfer and the Ethics of First-in-Human Research* (New York: Cambridge University Press, 2010), 26, 132.

34. "Wish-fulfilling medicine can be defined as 'doctors and other health professionals using medical means (medical technology, drugs, and so on) in a medical setting to fulfil the explicitly stated, prima facie non-medical wish of a patient.'" Eva C. A. Asscher, Ineke Bolt, and Maartje Schermer, "Wish-Fulfilling Medicine in Practice: A Qualitative Study of Physician Arguments," *Journal of Medical Ethics* 38 (2012): 327.

35. In Maxwell J. Mehlman, *Transhumanist Dreams and Dystopian Nightmares: The Promise and Peril of Genetic Engineering* (Baltimore: Johns Hopkins University Press, 2012), 5.

36. In Evans, *Playing God?*, 52.

37. Ramsey, *Fabricated Man*, 143–44.

38. Ronald Cole-Turner, "Religion and the Question of Human Germline Modification," in *Design and Destiny: Jewish and Christian Perspectives*

on *Human Germline Modification*, ed. Ronald Cole-Turner (Cambridge, MA: MIT Press, 2008), 1–27.
39. Evans, *Playing God?*.
40. W. French Anderson, "Prospects for Human Gene Therapy," *Science* 226 (1984): 402.
41. Theodore Friedmann, "Progress Toward Human Gene Therapy," *Science* 244 (1989): 1275.
42. LeRoy Walters, "Human Gene Therapy: Ethics and Public Policy," *Human Gene Therapy* 2 (1991): 117.
43. Anderson, "Prospects for Human Gene Therapy"; Friedmann, "Progress Toward Human Gene Therapy"; D. J. Weatherall, "The Slow Road to Gene Therapy," *Nature* 331 (1988): 13–14.
44. Recombinant DNA Advisory Committee Human Gene Therapy Subcommittee, "Points to Consider in the Design and Submission of Human Somatic Cell Gene Therapy Protocols," *Recombinant DNA Technical Bulletin* 8, no. 4 (1985): 182.
45. A. Dusty Miller, "Human Gene Therapy Comes of Age," *Nature* 357 (1992): 455–60.
46. Evans, *Playing God?*; John H. Evans, *The History and Future of Bioethics: A Sociological View* (New York: Oxford University Press, 2012).
47. S. E. Luria, "Directed Genetic Change: Perspectives from Molecular Genetics," in *The Control of Human Heredity and Evolution*, ed. T. M. Sonneborn (New York: Macmillan, 1965), 3, 17.
48. Tom L. Beauchamp and James F. Childress, *Principles of Biomedical Ethics* (New York: Oxford University Press, 1979). I recognize that in bioethics these values are called principles, and I have elsewhere described them as ends or goals. While these terms are subtly different in more theoretical writing, when it comes to application in actual debates, they are interchangeable, particularly at the high level of generalization at which I am operating in this book.
49. John C. Fletcher and W. French Anderson, "Germ-Line Gene Therapy: A New Stage of the Debate," *Law, Medicine and Health Care* 20 (1992): 31.
50. Walters, "The Ethics of Human Gene Therapy," 227.
51. Walters, "The Ethics of Human Gene Therapy," 227.
52. Friedmann, "Progress Toward Human Gene Therapy," 1280.
53. Professional ethicists will recognize the doctrine of double effect in this argument, that if doing good has a bad side effect, it's OK as long as you did not intend the bad side effect. See Alison McIntyre, "Doctrine of Double

Effect," *Stanford Encyclopedia of Philosophy* (2018) https://plato.stanford.edu/entries/double-effect/.
54. David J. Rothman, *Strangers by the Bedside: A History of How Law and Bioethics Transformed Medical Decision Making* (New York: Basic Books, 1991), 1–2.
55. Evans, *Playing God?*, 226.
56. John Robertson, "Genetic Alteration of Embryos: The Ethical Issues," in *Genetics and the Law III*, ed. Aubrey Milunsky and George Annas (New York: Plenum Press, 1985), 124–25.
57. Gregory Fowler, Eric T. Juengst, and Burke K. Zimmerman, "Germ-Line Therapy and the Clinical Ethos of Medical Genetics," *Theoretical Medicine* 10 (1989): 158, 159.
58. Walters, "Human Gene Therapy," 118.
59. Burke K. Zimmerman, "Human Germ-Line Therapy: The Case for Its Development and Use," *Journal of Medicine and Philosophy* 16 (1991): 594–95. See also Fletcher and Anderson, "Germ-Line Gene Therapy," 29.
60. Michael J. Sandel, "The Case Against Perfection," *The Atlantic Monthly*, April 2004, 5, 6.
61. Gilbert Meilaender, "Begetting and Cloning," *First Things*, June/July 1997, 41–43.
62. A barrier between selection and modification may have been successful not only because it largely fit with the values holding up the somatic/germline barrier, but also because it would have been supported by the "doing vs. allowing" distinction that has much weight in ethics, most often debated in terms of doing or allowing harm, and most often expressed in the distinction between killing and allowing to die. See Fiona Woollard, "Doing vs. Allowing Harm," *Stanford Encyclopedia of Philosophy* (2016), https://plato.stanford.edu/entries/doing-allowing/. In the case of HGE, if we look past all the "doing" in the process, "doing" would be modifying genes, and "allowing" would be selecting from genes that already are expressed in the embryos of the couple. Moreover, "an extensive body of research suggests that the distinction between doing and allowing plays a critical role in shaping moral appraisals" for the general public. See Fiery Cushman, Joshua Knobe, and Walter Sinnott-Armstrong, "Moral Appraisals Affect Doing/Allowing Judgments," *Cognition* 108 (2008): 281. We could then expect that the public would prefer a situation where an embryo with a particular genetic sequence was allowed to come into being rather than the situation where an embryo was modified to create that exact same genetic sequence. This would generally strengthen the barrier.

63. Fowler, Juengst, and Zimmerman, "Germ-Line Therapy and the Clinical Ethos of Medical Genetics," 157.
64. Juengst, "Can Enhancement Be Distinguished from Prevention in Genetic Medicine," 126.
65. W. French Anderson, "Human Gene Therapy: Why Draw a Line?" *Journal of Medicine and Philosophy* 14 (1989): 681–82, 687–88.
66. Fowler, Juengst, and Zimmerman, "Germ-Line Therapy and the Clinical Ethos of Medical Genetics," 160.
67. Fowler, Juengst, and Zimmerman, "Germ-Line Therapy and the Clinical Ethos of Medical Genetics," 160.
68. Fowler, Juengst, and Zimmerman, "Germ-Line Therapy and the Clinical Ethos of Medical Genetics," 161.
69. Walters was prescient. Noting that the techniques envisioned in 1986 would simply add a gene to try to make more of an enzyme, but that some diseases would require "gene replacement." "Techniques for this kind of precisely targeted molecular microsurgery have not yet been developed for mammals and will probably not emerge for decades." Walters, "The Ethics of Human Gene Therapy," 226, 227. They emerged in 2012.
70. President's Council on Bioethics, *Beyond Therapy: Biotechnology and the Pursuit of Happiness* (Washington, DC: President's Council on Bioethics, 2003), 13.
71. National Academies of Sciences, Engineering, and Medicine, *Human Genome Editing: Science, Ethics, and Governance* (Washington, DC: National Academies Press, 2017), 142.
72. Norman Daniels, "Normal Functioning and the Treatment-Enhancement Distinction," *Cambridge Quarterly of Healthcare Ethics* 9 (2000): 309.
73. Erik Parens, "Is Better Always Good? The Enhancement Project," in *Enhancing Human Traits: Ethical and Social Implications*, ed. Erik Parens (Washington, DC: Georgetown University Press, 1998), 7; Eric T. Juengst and Daniel Moseley, "Human Enhancement," *The Stanford Encyclopedia of Philosophy*, http://plato.stanford.edu/archives/sum2015/entries/enhancement 2015.
74. Evans, *The History and Future of Bioethics*.
75. "Gene therapy was first conceived as a treatment for hereditary single-gene defects." Clare E. Thomas, Anja Ehrhardt, and Mark A. Kay, "Progress and Problems with the Use of Viral Vectors for Gene Therapy," *Nature Reviews Genetics* 4 (2003): 347.
76. W. French Anderson, "Human Gene Therapy," *Science* 256 (1992): 808.

77. Juengst, "Can Enhancement Be Distinguished from Prevention in Genetic Medicine," 126.
78. Of the 100 protocols approved from 1990 to 1995, 22% were for genetic diseases, and 78% for cancers and other diseases not the result of genetic inheritance. Walters and Palmer, *The Ethics of Human Gene Therapy*, 25. A 2017 review shows that 2,597 trials had been undertaken in 38 countries, but that 11% of trials have been for genetic diseases, and thus 89% for other diseases. Cancer diseases are the largest category (65%), followed by infectious disease (7%) and cardiovascular disease (7%). Samantha L. Ginn et al., "Gene Therapy Clinical Trials Worldwide to 2017: An Update," *Journal of Gene Medicine* 20 (2018): 7.
79. P. R. Lowenstein, "Editorial," *Gene Therapy* 4 (1997): 755.
80. For example, Wakefield defines "harmful dysfunctions" as those that involve both a biological dysfunction and a cultural judgement that the dysfunction is harmful. Jerome C. Wakefield, "The Concept of Mental Disorder: On the Boundary Between Biological Facts and Social Values," *American Psychologist* 47, no. 3 (1992): 373–88. See more generally, Allan V. Horwitz, *What's Normal? Reconciling Biology and Culture* (New York: Oxford University Press, 2016).

Chapter 3

1. Clare E. Thomas, Anja Ehrhardt, and Mark A. Kay, "Progress and Problems with the Use of Viral Vectors for Gene Therapy," *Nature Reviews Genetics* 4 (2003): 346.
2. Luigi Naldini, "Gene Therapy Returns to Centre Stage," *Nature* 526, no. 351–60 (2015): 351.
3. Naldini, "Gene Therapy Returns to Centre Stage," 352.
4. Ray Noble et al., "Pandora's Box: Ethics of PGD for Inherited Risk of Late-Onset Disorders," *Ethics, Bioscience and Life* 3, no. 3 (2008).
5. Motoko Araki and Tetsuya Ishii, "International Regulatory Landscape and Integration of Corrective Genome Editing into in Vitro Fertilization," *Reproductive Biology and Endocrinology* 12, no. 108 (2014): 1–12.
6. Martin Jinek et al., "A Programmable Dual-RNA-Guided DNA Endonuclease in Adaptive Bacterial Immunity," *Science* (2012): 1225829.
7. Jennifer A. Doudna and Emmanuelle Charpentier, "The New Frontier of Genome Engineering with CRISPR-Cas9," *Science* 346, no. 6213 (2014): 1077; National Academies of Sciences, Engineering, and Medicine,

Human Genome Editing: Science, Ethics, and Governance (Washington, D.C: National Academies Press, 2017).

8. Doudna and Charpentier, "The New Frontier of Genome Engineering with CRISPR-Cas9," 1077.
9. David Cyranoski and Sara Reardon, "Chinese Scientists Genetically Modify Human Embryos," *Nature* 22 (April 2015).
10. Cyranoski and Reardon, "Chinese Scientists Genetically Modify Human Embryos."
11. NIH Director, "Statement on NIH Funding of Research Using Gene-Editing Technologies in Human Embryos," 2015.
12. At the time it was thought that men do not pass on their mitochondial DNA to their children. Moreover, the germline barrier was still central enough in these debates that the first charge given to the committee from the National Academies was to determine "whether manipulation of mitochondrial content should be considered germline modification . . . in the same way and with the same social and ethical implications as germline modification of nuclear DNA, or whether, from a social and ethical perspective, it should be viewed differently from germline modification of nuclear DNA." Another charge was to investigate whether to establish controls that "distinguish between genetic modification to prevent transmission of mitochondrial disease (therapeutic/prevention purposes) and genetic modification to enhance desired traits (enhancement purposes)." National Academies of Sciences, Engineering, and Medicine, *Mitochondrial Replacement Techniques: Ethical, Social, and Policy Considerations* (Washington, DC: National Academies Press, 2016), S-2.
13. Mosaicism is where some cells in an embryo (and later born human) have one set of genes and other cells have a different set of genes. One of the challenges in trying to change an embryo with CRISPR is that it cannot be determined if all of the cells are changed, and thus any born human may only have the change in some of the cells of their body—with unknown implications.Theodore Friedmann et al., "AsGCT and JSGT Joint Position Statement on Human Genomic Editing," *Molecular Therapy* 23, no. 8 (2015): 1282.
14. Edward Lanphier et al., "Don't Edit the Human Germ Line," *Nature* 519, no. 7544 (2015): 410–11.
15. Center for Genetics and Society, *Open Letter Calls for Prohibition on Reproductive Human Germline Modification*, Berkeley, CA, 2015..
16. Center for Genetics and Society, *Open Letter Calls for Prohibition on Reproductive Human Germline Modification*. The Center for Genetics

and Society released a joint report with Friends of the Earth in 2015 on advances in biology more generally. In the executive summary they discussed the various physical harms that could happen with germline. But, they pointed out that "some of those who are advocating for moratoria on editing the human germline nonetheless limit discussions of "ethics" to questions of scientific risk (safety), and fail to significantly consider social, ethical and legal risks." Their other defense of the germline was again as a traditional barrier on the slippery slope stopping the slide to Gattaca, given that the first applications downslope of the barrier would not create more inequality. Their argument is to point way downslope where social inequality lays: "The advent of human germline intervention could lead to the development of new forms of social inequality, discrimination and conflict. Among the risks of heritable genetic modification is the possibility of a modern version of eugenics, with human society being divided into genetic 'haves' and 'have-nots.'" Center for Genetics and Society and Friends of the Earth, *Extreme Genetic Engineering and the Human Future* (Berkeley: Center for Genetics and Society, 2015), 6.
17. UNESCO, "UNESCO Panel of Experts Calls for Ban on 'Editing' of Human DNA to Avoid Unethical Tampering with Hereditary Traits," 2015.
18. Nationale Akademie der Wissenschaften Leopoldina, "The Opportunities and Limits of Genome Editing," 2015, 26.
19. I am not the only person to see how few and narrow are the values in this debate. A recent project of the Hastings Center sought to identify the "non-physical harms" in gene editing because these were otherwise being ignored in the debate. See Erik Parens and Josephine Johnston, *Human Flourishing in an Age of Gene Editing* (New York: Oxford University Press, 2019).
20. The Asilomar debates concerned the safety of the first genetically modified organisms, and earned their title from the name of the conference center where a meeting was held in 1975 (Hurlbut 2015).
21. David Baltimore et al., "A Prudent Path Forward for Genomic Engineering and Germline Gene Modification," *Science* 348, no. 6230 (2015): 36–38.
22. "On Human Gene Editing: International Summit Statement," distributed at summit, December 3, 2015.
23. I was one of 22 members. While these NASEM documents are called consensus documents, obviously the consensus is that the final product should be released, not that every member agrees with every aspect of the report. What I write here is my own opinion of the HGE debate and the NASEM report. This book does not represent the views of the NASEM or

the other committee members. I do not use the confidential deliberations of the committee as data but rather only the published report, except for the following fact. In this chapter, I describe barriers that can be built from the NASEM report. These barriers were not rejected by the group, but rather I failed to consider the possibility of these barriers until long after the NASEM report was completed.

24. John H. Evans, *Playing God? Human Genetic Engineering and the Rationalization of Public Bioethical Debate* (Chicago: University of Chicago Press, 2002), 135.
25. The National Academies of Science (NAS) was established in 1863 by an Act of Congress as a private, non-governmental institution to advise the nation on issues related to science and technology. The National Academy of Engineering and the National Academy of Medicine were later established under the charter of the NAS and now work together. The Academies "provide independent, objective analysis and advice to the nation and conduct other activities to solve complex problems and inform public policy decisions"; see their website at http://www.nationalacademies.org/.
26. Evans, *Playing God?*, 135–38.
27. Antonio Regalado, "Rogue Chinese CRISPR Scientist Cited US Report as His Green Light," *Technology Review*, November 27, 2018.
28. "Statement by the Organizing Committee of the Second International Summit on Human Genome Editing," November 28, 2018, https://www.nationalacademies.org/news/2018/11/statement-by-the-organizing-committee-of-the-second-international-summit-on-human-genome-editing
29. These are "maximum transparency consistent with patient privacy continued reassessment of both health and societal benefits and risks, with broad ongoing participation and input by the public; and reliable oversight mechanisms to prevent extension to uses other than preventing a serious disease or condition." National Academies of Sciences, Engineering, and Medicine (hereafter in notes NASEM) *Human Genome Editing: Science, Ethics, and Governance* (Washington, DC: National Academies Press, 2017), 7–8.
30. These are "availability of credible preclinical and/or clinical data on risks and potential health benefits of the procedures; ongoing, rigorous oversight during clinical trials of the effects of the procedure on the health and safety of the research participants; comprehensive plans for long-term, multigenerational follow-up that still respect personal autonomy." NASEM, *Human Genome Editing*, 7–8.

31. NASEM, *Human Genome Editing*, 7–8.
32. NASEM, *Human Genome Editing*, 6.
33. NASEM, *Human Genome Editing*, 111.
34. "Heritable genome editing is not the only way to accomplish this goal" they write. "Other options include deciding not to have children; adopting a baby; or using donated embryos, eggs, or sperm. These options, however, do not allow both parents to have a genetic connection to their children, which is of great importance to many people." NASEM, *Human Genome Editing*, 113. The leaders of the NASEM committee continue to try to make this subtle point clear to audiences. See Prashant Nair, "QnAs with Alta Charo and George Church," *PNAS* 114, no. 23 (2017): 5769–71; Richard O. Hynes, Barry S. Coller, and Matthew Porteus, "Toward Responsible Human Genome Editing," *JAMA* 317, no. 18 (2017): 1829–30. It does not seem to have been noticed by people responding to the report. See Peter Sykora and Arthur Caplan, "The Council of Europe Should Not Reaffirm the Ban on Germline Genome Editing in Humans," *EMBO Reports* 18, no. 11 (2017): 1871–72; Eric T. Juengst, "Crowdsourcing the Moral Limits of Human Gene Editing?" *Hastings Center Report* 47, no. 3 (2017): 15–23.
35. In supporting the point that many parents care deeply about genetic relatedness, the report states that "the desire for genetic relation is evidenced by the fact that many prospective parents, faced with the choice between foregoing genetically related children or risking the birth of a child with a genetic illness, will choose to risk having an affected child." NASEM, *Human Genome Editing*, 120. They could have followed up on this and argued the HGE is beneficent to the children of parents who value genetic relatedness so much that they put their children at physical risk of having genetic disease. They did not.
36. NASEM, *Human Genome Editing*, 120.
37. "In other populations, the frequency of particular disease-causing mutations may be high enough that there is a significant chance that both prospective parents will be carriers of mutations in the same gene. Examples include the tumor suppressor genes BRCA1 and BRCA2, which increase the risk of breast and ovarian cancer even when inherited in a single copy (because of loss of the unaffected copy of the gene) and Tay-Sachs disease and other early-onset lysosomal storage diseases that are caused by the inheritance of two copies of recessive mutations." NASEM, *Human Genome Editing*, 114.
38. "As the survival of people with severe recessive diseases like cystic fibrosis, sickle-cell anemia, thalassemia, and lysosomal storage diseases improves

with advances in medical treatments, the possibility cannot be dismissed that there will be an increase in the number of situations in which both prospective parents are homozygous for a mutation. The societal and medical pressures faced by these people often bring them together in social groups where they are more likely to interact and develop close relationships. Similar associations can develop among patients with autosomal dominant genetic diseases that allow development to reproductive age (e.g., achondroplasia, osteogenesis imperfecta), again increasing the likelihood of transmitting disease alleles. As our ability to treat children and adults with serious genetic diseases improves . . . there may be a growing need to address concerns potential parents might have about passing along these diseases to their children." NASEM, *Human Genome Editing*, 114.
39. NASEM, *Human Genome Editing*, 115.
40. NASEM, *Human Genome Editing*, 115.
41. Eric Lander of the Broad Institute of Harvard and MIT pointed out that for recessive diseases, with PGD 75% of embryos would not have the disease, so finding an embryo to implant is generally not a problem. For the more rare dominant diseases, when one parent is heterozygous (the most common situation), 50% of embryos would be free of disease. (Lander presentation, 19:58 minute mark.) His conclusion is that almost all such couples can use PGD, although there may be some people who have bad luck in only getting embryos with disease. This would be the case when one parent is homozygous for a dominant disease or both parents are homozygous for a recessive disease. But, as he wrote elsewhere, "such situations are vanishingly rare for most monogenic diseases. For dominant Huntington's disease, for example, the total number of homozygous patients in the medical literature is measured in dozens. For most recessive disorders, cases are so infrequent (1 per 10,000 to 1 per million) that marriages between two affected persons will hardly ever occur unless the two are brought together by the disorder itself." See Eric S. Lander, "Brave New Genome," *New England Journal of Medicine* 373, no. 1 (2015): 6.
42. NASEM, *Human Genome Editing*, 113. This point has been made by many individuals in the HGE debate. For example, Sykora and Caplan state that "for some, genetic engineering in fact provides a more ethical option as it does not involve the destruction of embryos when it is done on germ cells or stem-cell precursors of germ cells." Sykora and Caplan, "The Council of Europe Should Not Reaffirm the Ban on Germline Genome Editing in Humans," 1871. Biologist George Church asks in the *New England*

Journal of Medicine whether "we want to lose the moral high ground on what would be a way of reducing abortions and losses of embryos?" George Church, "Compelling Reasons for Repairing Human Germlines," *New England Journal of Medicine* 377, no. 20 (2017): 1910. In the words of a group of French scholars, we may "have a moral duty to cure affected human embryos instead of discarding them." Julie Steffann et al., "Could Failure in Preimplantation Genetic Diagnosis Justify Editing the Human Embryo Genome?" *Cell Stem Cell* no. 22 (2018): 481. To make a massive generalization in the American context, the non-religious, liberal Protestants, liberal Catholics, and most religious minorities tend to be less concerned about embryonic life, and conservative Protestants and traditionalist Catholics are more concerned.

43. Paul Scherz, "The Mechanism and Applications of CRISPR-Cas9," *The National Catholics Bioethics Quarterly* Spring (2017): 34.
44. Scherz, "The Mechanism and Applications of CRISPR-Cas9," 34. George Church writes that "in 2004, a Vatican commission stated that 'Germ line genetic engineering with a therapeutic goal in man would in itself be acceptable . . . in the stem cells that produce a man's sperm, whereby he can beget healthy offspring with his own seed by means of the conjugal act.'" Church, "Compelling Reasons for Repairing Human Germlines," 1910.
45. Juengst, "Crowdsourcing the Moral Limits of Human Gene Editing?"
46. NASEM, *Human Genome Editing*, 147, 148.
47. Indeed, there are chapters of the report dedicated to the traditional barriers in the debate—one on "somatic genome editing," one on "heritable genome editing," as well as one titled "enhancement." There is not a chapter titled "disease."
48. Elsewhere it is defined as "changes that alter what is 'normal,' whether for humans as a whole or for a particular individual prior to enhancement." They also review various definitions that typically concern "improvement" and being unusual. For example, they write that "enhancement has been variously defined as 'boosting our capabilities beyond the species-typical level or statistically normal range of functioning,' . . . 'a nontherapeutic intervention intended to improve or extend a human trait,' . . . or 'improvements in the capacities of existing individuals or future generations.'" Another definition they identify "focuses on interventions that improve bodily condition or function beyond what is needed to restore or sustain health." NASEM, *Human Genome Editing,* 137, 138, 145.
49. NASEM, *Human Genome Editing*, 148–49.
50. NASEM, *Human Genome Editing*, 123.

51. Allen Buchanan et al., *From Chance to Choice: Genetics and Justice* (New York: Cambridge University Press, 2000), 81.
52. NASEM, *Human Genome Editing*, 150.
53. NASEM, *Human Genome Editing*, 7. My emphasis. This condition actually ends with "with little or no evidence of adverse effects." This second clause is simply another call for only conducting germline HGE that is safe.
54. I am using trait in a very micro way. Each variant produces a unique trait in the human body, even if these traits result in the same observable outcome in the human body.
55. See the report on standards and guidelines for the interpretation of sequence variants by the American College of Medical Genetics and Genomics and the Association of Molecular Pathology. When a medical provider sequences a patient's gene sequence at the location of known disorders, like BRCA1 for cancer, the laboratory classifies the variant as pathogenic, likely pathogenic, uncertain significance, likely benign, or benign. A vast majority of identified variants are benign, likely benign, or of uncertain significance. The report states that "assessing the frequency of a variant in a control or general population is useful in assessing its potential pathogenicity" and that a frequency of over 5% is considered to be "stand alone support" for a variant being benign for a rare Mendelian disorder. Sue Richards et al., "Standards and Guidelines for the Interpretation of Sequence Variants: A Joint Consensus Recommendation of the American College of Medical Genetics and Genomics and the Association for Molecular Pathology," *Genetics in Medicine* 17, no. 5 (2015): 10.
56. Jackie Leach Scully, "Choice, Chance, and Acceptance," in *Human Flourishing in an Age of Gene Editing*, ed. Erik Parens and Josephine Johnston (New York: Oxford University Press, 2019), 143–56; Rosemarie Garland-Thomson, "Welcoming the Unexpected," in *Human Flourishing in an Age of Gene Editing*, ed. Erik Parens and Josephine Johnston (New York: Oxford University Press, 2019), 15–28.
57. "It is also possible, however, to envision the possibility of changing a gene to a variant form that does not exist (or is rare) in the human gene pool but has some property that could be viewed as an 'enhancement' since it is predicted to have a beneficial effect. Such a change is a more radical step than that of replacing a disease-causing variant with a common human variant known not to cause disease." NASEM, *Human Genome Editing*, 139.
58. Google dictionary.
59. Dictionary.com.

60. There is a feature of continuity vagueness slippery slopes that has not been discussed much in this literature, which is that it is much more possible to create a compromise that can hold with continuity instead of similarity vagueness. With similarity vagueness, which affects the disease/enhancement distinction, it is not clear why one would draw a line between susceptibility to obesity vs. susceptibility to mental illness. With continuity vagueness, the inability to set small differences is acknowledged while pointing to the end points of the scale for which there is consensus. An arbitrary compromise is then required and can be enforced as such. For example, no one thinks that the average 9-year-old is capable of driving, and everyone thinks that the average 21-year-old is capable of driving. In the United States, age 16 is an arbitrary barrier, selected as a compromise, and enforced, but nobody advocates setting the number at 9 or 25. The reason practical compromise does not work for the gene editing debate is that, to continue my metaphor, some would want it set at 9 (or less) and some at 25 (or more).

61. If this barrier were eventually made into policy, regulators might look for the meaning of "prevalent" in related policies or institutionalized meanings of the term in medicine and science. "Rare" is an antonym of prevalent, and "rare disease" has been statutorily defined in many countries, including the United States and the European Union. For example, in the European Union a "rare disease" is one found in less than 1 in 2,000 people (5/100ths of 1%). In the United States "rare" means "effects <200,000 people." When that was set by the government in 1984, this meant rare was at 8/100ths of 1%. With population growth, today this would mean 6/100ths of 1%. A study that searched for definitions across world organizations found an average definition of rare disease of 1 in 2,500. See Trevor Richter et al., "Rare Disease Terminology and Definitions—A Systematic Global Review: Report of the ISPOR Rare Disease Special Interest Group," *Value in Health* 18 (2015): 906–14.

If "prevalent" was then taken to mean "not rare," by this definition people could look through the human population for advantageous variants found in greater than 5/100ths of 1% of the population. While I do not think geneticists have reached any conclusions, there may well be advantageous variants that are found in 6/100ths of 1% of the population, which would make this definition not fit with the value of fairness as defined earlier.

There are other uses of the word "common" in genetic science, but there does not seem to be agreement on the meaning of the term. For example, a recent review of GWAS states that they are going to "arbitrarily" make a definition where "common variants" are those found in greater than 1% of the population and "rare variants" are those found in less than 1%. See Peter M. Visscher et al., "10 Years of GWAS Discovery: Biology, Function, and Translation," *American Journal of Human Genetics* 101 (2017): 6. If the barrier was built on this definition, and he was limited to installing variants found in 1% of the population, this would certainly give less opportunity for producing social advantage than the 5/100ths of 1% criteria suggested by the European Union definition of rare disease but could still provide traits that were socially advantageous.

62. Centers for Disease Control, "Malaria," https://www.cdc.gov/parasites/malaria/index.html. Additionally, at present, most population databases in wealthy Western countries reflect the racial makeup of the advantaged population, which can be described as "Euro-centric data sets." However, there are already racially and geographically diverse databases that can be searched. For example, the recently published standards for defining gene variants suggests searching three databases. One is the 1000 Genomes Project, which is a compendium of analyses of 2,504 genomes from 26 geographically dispersed populations from five continental regions. (As expected, the project concludes that given human origins in Africa, and that a very small group of humans apparently left Africa to be the ancestors of everyone else, "most of the world's variation between humans occurs in sub-Saharan populations." Ewan Birney and Nicole Soranzo, "The End of the Start for Population Sequencing," *Nature* 526 (2015): 53. I assume that by the time any of this technology is possible, which will be many years from now, these gene-sequence databases will contain more and more data and be more and more representative of the population in which they are based.

63. Population genetics results show that "variants present at 10% and above across the entire sample are almost all found in all of the populations studied. By contrast, 17% of low-frequency variants in the range 0.5–5% were observed in a single ancestry group, and 53% of rare variants at 0.5% were observed in a single population." 1000 Genomes Project Consortium, "An Integrated Map of Genetic Variation from 1,092 Human Genomes," *Nature* 491 (2012): 58.

64. In Paul Ramsey, *Fabricated Man: The Ethics of Genetic Control* (New Haven, CT: Yale University Press, 1970), 1.

65. Andrew et al. Wood, "Defining the Role of Common Variation in the Genomic and Biological Architecture of Adult Human Height," *Nature Genetics* 46, no. 11 (2014): 1173.

66. Dalton Conley and Jason Fletcher, *The Genome Factor: What the Social Genomics Revolution Reveals About Ourselves, Our History, and the Future* (Princeton, NJ: Princeton University Press, 2017), 43ff.
67. Visscher et al., "10 Years of GWAS Discovery," 8.
68. Conley and Fletcher, *The Genome Factor*, 55–56, 171.
69. Conley and Fletcher, *The Genome Factor*, 172. For a list of technical requirements needed for this technology to come to pass, see Henry T. Greely, *The End of Sex and the Future of Human Reproduction* (Cambridge, MA: Harvard University Press, 2016), Ch. 7.
70. Daniel J. Schaid, Wenan Chen, and Nicholas B. Larson, "From Genome-Wide Associations to Candidate Causal Variants by Statistical Fine-Mapping," *Nature Reviews Genetics* 19 (2018): 491–504.
71. In selection via PGD, we know that the variants produce a functioning human because the parents produced them. To avoid modifications that produce combinations that have not existed in humans before, parents could modify the variants to make the baby as close to the median actual existing human in the database used to create the scale. For non-continuous scales like "educational attainment," there will be many different combinations that could be used to produce a predicted median score of 16 years, for example.
72. People could argue for the use of mean (average), which is different than the median by incorporating information about the magnitude of difference between people. For example, in a population of three people with educational attainment of 8, 12, and 16 years, the median is 12 and the mean is 12. But, if the final person above did not have 16 but rather 19 years of education (perhaps they went to law school), the median would still be 12, but the mean would be 13. If you were allowed to modify your embryo to give it the mean of 13, it would no longer be exactly between advantage and disadvantage, but would have an advantage over the bottom 50% of the people in the distribution (i.e., below the median of 12). This is inconsistent with fairness as I describe it.

    Mode is the most common value in a distribution but would be an even worse fit with fairness. Imagine a fairly flat distribution of a scale from 1 to 100, where each value has two people in it, except for 27, which has 3. The mode would be 27 and the median 50. This would mean that people at 34 could not use HGE to "level the playing field" but could only make themselves worse. This also does not fit with the overarching value of fairness.
73. This objectively defined barrier for the polygenic context would, like the most prevalent variant barrier, not define or give a valuation for traits and

rely upon the biological fact that more prevalent variants will tend to result in a more ordinary body. That is, we would revert to the most prevalent variant barrier I identified earlier but applied to as many genes as a person wants. Specifically, imagine a couple that is aware of the variants that comprise different levels of the polygenic educational attainment score. They could use gene editing to change any one or more of those genes in the score to the most prevalent version in the population. Similarly, they could use PGD to select the embryo that had the most prevalent scores averaged across all the genes tested in the embryo. This barrier would not be susceptible to either continuity or similarity vagueness because there is only one most prevalent variant of each gene in a population and, unlike the median trait barrier, there is no consideration of what is valued as higher and lower which the median trait measure depends upon. This barrier would be defined only by the objective fact of the distribution of variants in the population.

What this barrier gains in design strength, it may lose in a fuzzier connection to the value of fairness, resulting in cases labeled as "fair" being on both sides of the barrier. The size of the trade-off between the median trait and this more objective barrier depends on how far selecting the most prevalent version of every gene in a polygenic score puts you from the median of the polygenic risk score. There is a gap because genes are not additive, with each resulting in a 1% "improvement," (for example) but rather are highly interactive and non-linear in ways that nobody understands.

For example, imagine that people whose children would be disadvantaged in educational attainment want to use the polygenic scale with a median of 50 to level the playing field for their children. They would otherwise produce a child with a 30, but changing each of the genes in the scale to the most prevalent variant only gets them to 40 on the scale—not a perfectly level playing field. That said, presumably polygenic scores found toward the ends of the continuum are unusual and therefore are caused by unusual gene variants. Thus, allowing people to modify or select the most prevalent of any gene they wanted would generally push outcomes toward the middle of the scale, resulting in a general tendency to level the playing field and the inability to gain social advantage. Whether people in some future debate would want to rely only on the more solid "polygenic most prevalent variant" barrier will depend upon how close the most prevalent variants of all of the components of scales that people want are to the median outcome on a trait. To my knowledge, this is all quite unknown.

A final possible definition of this barrier suffers from the same problem. We could identify the most common combination of variants in the population used to make the polygenic scale, identifying the genetically median human on that trait. It could be that 10% of all people in the population have the exact same combination of 100 variants that predict height, and this combination could be considered the most prevalent. For example, imagine that there are only three genes in a polygenic score, and that each gene has only three variants: A, B, C for the first; D, E, F for the second; and G, H, I for the third. Person 1 is A, D, I. Person 2 is A, E, I. Person 3 is A, D, I. Person 4 is A, D, H and Person 5 is A, D, I. If there were only these five people in the population, the most prevalent combination would be A, D, I. This would be the median human on the genes measured by the score. This basic example could be expanded to hundreds of genes with hundreds of variants for each for a million people to make a population with which to calculate the most prevalent combination. This version would not be as tight a fit with the value of fairness because what we could call "the median human" in variants may not have the median trait, which is what matters for social outcomes.

74. Herbert McClosky and John Zaller, *The American Ethos: Public Attitudes Toward Capitalism and Democracy* (Cambridge, MA: Harvard University Press, 1984), 66, 83.
75. Neil Gross, *Why Are Professors Liberal and Why Do Conservatives Care?* (Cambridge, MA: Harvard University Press, 2013).
76. "On Human Gene Editing: International Summit Statement."
77. Buchanan et al., *From Chance to Choice*.
78. Buchanan et al., *From Chance to Choice*, 64, 81.
79. Buchanan et al., *From Chance to Choice*, 82.
80. Center for Genetics and Society, *Open Letter*.
81. NASEM, *Human Genome Editing*, 150.
82. Lander, "Brave New Genome," 6–7.
83. Patrick Prag and Melinda C. Mills, "Assisted Reproductive Technology in Europe: Usage and Regulation in the Context of Cross-Border Reproductive Care," in *Childlessness in Europe: Contexts, Causes, and Consequences*, ed. Michaela Kreyenfeld and Dirk Konietzka, 289–309 (Cham, Switzerland: Springer, 2017), 296.
84. Andrew P. Wilper et al., "Health Insurance and Morality in US Adults," *American Journal of Public Health* 99, no. 12 (2009): 2289, 2294.
85. Sara J. Solnick and David Hemenway, "Is More Always Better?: A Survey on Positional Concerns," *Journal of Economics Behavior and Organization* 37 (1998): 373–83.

86. John Harris, *Enhancing Evolution: The Ethical Case for Making Better People* (Princeton, NJ: Princeton University Press, 2007), 62.
87. Julian Savulescu, "Procreative Beneficence: Why We Should Select the Best Children," *Bioethics* 15, nos. 5–6 (2001): 425.
88. John A. Robertson, "Procreative Liberty in the Era of Genomics," *American Journal of Law and Medicine* 29 (2003): 444, 445.

Chapter 4

1. Marilynn Marchione, "Chinese Researcher Claims First Gene-Edited Babies," Associated Press, November 26, 2018.
2. John H. Evans, *The History and Future of Bioethics: A Sociological View* (New York: Oxford University Press, 2012).
3. Erik Parens, "Is Better Always Good? The Enhancement Project," in *Enhancing Human Traits: Ethical and Social Implications*, ed. Erik Parens (Washington, DC: Georgetown University Press, 1998), S3.
4. Eric T. Juengst, "Can Enhancement Be Distinguished from Prevention in Genetic Medicine," *Journal of Medicine and Philosophy* 22 (1997): 131.
5. Eric T. Juengst and Daniel Moseley, "Human Enhancement," *The Stanford Encyclopedia of Philosophy*, http://plato.stanford.edu/archives/sum2015/entries/enhancement 2015.
6. Juengst and Moseley, "Human Enhancement."
7. Peter Conrad, *The Medicalization of Society: On the Transformation of Human Conditions into Treatable Disorders* (Baltimore, MD: Johns Hopkins University Press, 2007).
8. Juengst and Moseley, "Human Enhancement."
9. Nicolas Rivron et al., "Debate Ethics of Embryo Models from Stem Cells," *Nature* 564 (2018): 183–85.
10. Henry T. Greely, *The End of Sex and the Future of Human Reproduction* (Cambridge, MA: Harvard University Press, 2016), Ch. 8.
11. Sonia M. Suter, "The Tyranny of Choice: Reproductive Selection in the Future," *Journal of Law and Biosciences* 5 (2018): 277, 287ff. On pages 278 to 282 the author also gives a nice summary of what would provide the grease on the slippery slope toward expansion of this technology, such as pressure from insurance companies, claims about "responsible parenthood," legal concerns of fertility clinics, cultural notions that choice and control are always better, and much more.
12. In a 2009 article I reported that the website of the Association of Reproductive Health Professionals had defined enhancement as

when "a couple would like to endow their child with genes that neither member of the couple possesses."John H. Evans and Cynthia E. Schairer, "Bioethics and Human Genetic Engineering," in *Handbook of Genetics and Society: Mapping the New Genomic Era*, ed. Paul Atkinson, Peter Glasner, and Margaret Lock (London: Routledge, 2009), 355. This definition no longer exists on their website.
13. The subtle difference is that users of WGS-PGD can select from among embryos they produce after the probabilistic combination of the genes of the two parents, and HGE would allow for a more expansive list of any traits they *could* produce, given unlimited tries. At this point of our technological development, genetic scientists should presumably be able to determine exactly which combinations on a polygenic index a couple is capable of producing.
14. If a population geneticist were to conclude that the odds of this scenario are not essentially zero, the definition of this barrier could be expanded to the eight grandparents of the prospective parents with little impact on the outcome.
15. Eric S. Lander, "Brave New Genome," *New England Journal of Medicine* 373, no. 1 (2015): 7.
16. Lander, "Brave New Genome," 7.
17. Nicholas Agar, *Humanity's End: Why We Should Reject Radical Enhancement* (Cambridge, MA: MIT Press, 2010), 17, 1.
18. Helen de Cruz and Johan de Smedt, "The Role of Intuitive Ontologies in Scientific Understanding—The Case of Human Evolution," *Biology and Philosophy* 22 (2006): 352.
19. Cruz and Smedt, "The Role of Intuitive Ontologies in Scientific Understanding," 353.
20. To really get into the details, the public who uses this idea tends to think that "made by" means created by God, one by one, and this barrier would be inconsistent with this version of the value. Academics tend to think that "made by" is more like "designed by," which would be more consistent with this value. John H. Evans, *What Is a Human? What the Answers Mean for Human Rights* (New York: Oxford University Press, 2016), Ch. 2, 6.
21. Anna L. Peterson, *Being Human: Ethics, Environment, and Our Place in the World* (Berkeley: University of California Press, 2001).
22. Evans, *The History and Future of Bioethics*.
23. John Harris, *Enhancing Evolution: The Ethical Case for Making Better People* (Princeton, NJ: Princeton University Press, 2007), 79.

24. John A. Robertson, "Procreative Liberty in the Era of Genomics," *American Journal of Law and Medicine* 29 (2003): 445, 448.
25. Daniel Kevles, *In the Name of Eugenics: Genetics and the Uses of Human Heredity* (Berkeley: University of California Press, 1985), 164ff.
26. Allen Buchanan et al., *From Chance to Choice: Genetics and Justice* (New York: Cambridge University Press, 2000), Ch. 6.
27. Buchanan et al., *From Chance to Choice*, 170–72.
28. National Academies of Sciences, Engineering, and Medicine (NASEM), *Human Genome Editing: Science, Ethics, and Governance* (Washington, DC: National Academies Press, 2017), 119.
29. NASEM, *Human Genome Editing*, 121–30.
30. Evans, *The History and Future of Bioethics*.

Chapter 5

1. Eric T. Juengst, "Crowdsourcing the Moral Limits of Human Gene Editing?" *Hastings Center Report* 47, no. 3 (2017): 21.
2. Juengst, "Crowdsourcing the Moral Limits of Human Gene Editing?," 22.
3. Sheila Jasanoff, *Designs on Nature: Science and Democracy in Europe and the United States* (Princeton, NJ: Princeton University Press, 2005).
4. Theodore M. Porter, *Trust in Numbers: The Pursuit of Objectivity in Science and Public Life* (Princeton, NJ: Princeton University Press, 1995), 194–95.
5. Human Fertilisation & Embryology Authority, "Approved PGD and PTT conditions." https://www.hfea.gov.uk/treatments/embryo-testing-and-treatments-for-disease/approved-pgd-and-ptt-conditions/. See elsewhere on their web pages for the list of approved genetic traits that can be avoided with PGD.
6. Nick Bostrom, "Letter from Utopia," *Studies in Ethics, Law and Technology* 2, no. 1 (2008): 1–7.

Bibliography

1000 Genomes Project Consortium. 2012. An Integrated Map of Genetic Variation from 1,092 Human Genomes. *Nature* 491: 56–65.
Abbott, Andrew. 1988. *The System of Professions: An Essay on the Division of Expert Labor*. Chicago: University of Chicago Press.
Agar, Nicholas. 2010. *Humanity's End: Why We Should Reject Radical Enhancement*. Cambridge, MA: MIT Press.
Almendraia, Anna. 2018. World's First Gene-Edited Babies Could Set Genetic Science Backward, Experts Worry. *Huffington Post*. November 27.
Anderson, W. French. 1972. Genetic Therapy. In *The New Genetics and the Future of Man*, edited by Michael Hamilton, 109–24. Grand Rapids, MI: William B. Eerdmans.
Anderson, W. French. 1984. Prospects for Human Gene Therapy. *Science* 226: 401–09.
Anderson, W. French. 1989. Human Gene Therapy: Why Draw a Line? *Journal of Medicine and Philosophy* 14: 681–93.
Anderson, W. French. 1990. Genetics and Human Malleability. *Hastings Center Report* 20, no. 1: 21–24.
Anderson, W. French. 1992. Human Gene Therapy. *Science* 256: 808–13.
Anderson, W. French, and John C. Fletcher. 1980. Gene Therapy in Human Beings: When Is It Ethical to Begin? *New England Journal of Medicine* 303: 1293–300.
Aposhian, H. Vasken. 1970. The Use of DNA for Gene Therapy—The Need, Experimental Approach, and Implications. *Perspectives in Biology and Medicine* 14: 98–108.
Araki, Motoko, and Tetsuya Ishii. 2014. International Regulatory Landscape and Integration of Corrective Genome Editing into In Vitro Fertilization. *Reproductive Biology and Endocrinology* 12, no. 108: 1–12.
Asscher, Eva C. A., Ineke Bolt, and Maartje Schermer. 2012. Wish-Fulfilling Medicine in Practice: A Qualitative Study of Physician Arguments. *Journal of Medical Ethics* 38: 327–31.
Baltimore, David, Paul Berg, Michael Botchan, Dana Carroll, R. Alta Charo, George Church, Jacob E. Corn, George Q. Daley, Jennifer A. Doudna, Marsha Fenner, Henry T. Greely, Martin Jinek, G. Steven Martin, Edward Penhoet, Jennifer Puck, Samuel H. Sternberg, Jonathan S. Weissman,

and Keith R. Yamamoto. 2015. A Prudent Path Forward for Genomic Engineering and Germline Gene Modification. *Science* 348, no. 6230: 36–38.

Beauchamp, Tom L., and James F. Childress. 1979. *Principles of Biomedical Ethics*. New York: Oxford University Press.

Beauchamp, Tom L., and James F. Childress. 2013. *Principles of Biomedical Ethics*. 7th ed. New York: Oxford University Press.

Birney, Ewan, and Nicole Soranzo. 2015. The End of the Start for Population Sequencing. *Nature* 526: 52–53.

Bok, Sissela. 1971. The Leading Edge of the Wedge. *Hastings Center Report* 1, no. 3: 9–11.

Bostrom, Nick. 2008. Letter from Utopia. *Studies in Ethics, Law and Technology* 2, no. 1: 1–7.

Buchanan, Allen, Dan W. Brock, Norman Daniels, and Daniel Wikler. 2000. *From Chance to Choice: Genetics and Justice*. New York: Cambridge University Press.

Center for Genetics and Society. 2015. *Open Letter Calls for Prohibition on Reproductive Human Germline Modification*. Berkeley, CA: Center for Genetics and Society.

Center for Genetics and Society and Friends of the Earth. 2015. *Extreme Genetic Engineering and the Human Future*. Berkeley, CA: Center for Genetics and Society.

Church, George. 2017. Compelling Reasons for Repairing Human Germlines. *New England Journal of Medicine* 377, no. 20: 1909–11.

Cole-Turner, Ronald. 2008. Religion and the Question of Human Germline Modification. In *Design and Destiny: Jewish and Christian Perspectives on Human Germline Modification*, edited by Ronald Cole-Turner, 1–27. Cambridge, MA: MIT Press.

Conley, Dalton, and Jason Fletcher. 2017. *The Genome Factor: What the Social Genomics Revolution Reveals About Ourselves, Our History, and the Future*. Princeton, NJ: Princeton University Press.

Conrad, Peter. 2007. *The Medicalization of Society: On the Transformation of Human Conditions into Treatable Disorders*. Baltimore, MD: Johns Hopkins University Press.

Corner, Adam, Ulrike Hahn, and Mike Oaksford. 2011. The Psychological Mechanism of the Slippery Slope Argument. *Journal of Memory and Language* 64: 133–52.

Cruz, Helen de, and Johan de Smedt. 2006. The Role of Intuitive Ontologies in Scientific Understanding—The Case of Human Evolution. *Biology and Philosophy* 22: 351–68.

Cushman, Fiery, Joshua Knobe, and Walter Sinnott-Armstrong. 2008. Moral Appraisals Affect Doing/Allowing Judgments. *Cognition* 108: 281–89.

Cyranoski, David, and Sara Reardon. 2015. Chinese Scientists Genetically Modify Human Embryos. *Nature,* April 22.

Daniels, Norman. 2000. Normal Functioning and the Treatment-Enhancement Distinction. *Cambridge Quarterly of Healthcare Ethics* 9: 309–22.

Davis, Bernard D. 1970. Prospects for Genetic Intervention in Man. *Science* 170: 1279–83.

Doudna, Jennifer A., and Emmanuelle Charpentier. 2014. The New Frontier of Genome Engineering with CRISPR-Cas9. *Science* 346, no. 6213: 1077.

Evans, John H. 2002. *Playing God? Human Genetic Engineering and the Rationalization of Public Bioethical Debate.* Chicago: University of Chicago Press.

Evans, John H. 2012. *The History and Future of Bioethics: A Sociological View.* New York: Oxford University Press.

Evans, John H. 2016. *What Is a Human? What the Answers Mean for Human Rights.* New York: Oxford University Press.

Evans, John H., and Cynthia E. Schairer. 2009. Bioethics and Human Genetic Engineering. In *Handbook of Genetics and Society: Mapping the New Genomic Era,* edited by Paul Atkinson, Peter Glasner, and Margaret Lock, 349–66. London: Routledge.

Fletcher, John C., and W. French Anderson. 1992. Germ-Line Gene Therapy: A New Stage of the Debate. *Law, Medicine and Health Care* 20: 26–39.

Fowler, Gregory, Eric T. Juengst, and Burke K. Zimmerman. 1989. Germ-Line Therapy and the Clinical Ethos of Medical Genetics. *Theoretical Medicine* 10: 151–65.

Friedmann, Theodore. 1989. Progress Toward Human Gene Therapy. *Science* 244: 1275–81.

Friedmann, Theodore, and Richard Roblin. 1972. Gene Therapy for Human Genetic Disease? *Science* 175: 949–55.

Friedmann, Theodore, Erica C. Jonlin, Nancy MP King, Bruce E. Torgett, Nelson A. Wivel, Yasufumi Kaneda, and Michael Sadelain. 2015. AsGCT and JSGT Joint Position Statement on Human Genomic Editing. *Molecular Therapy* 23, no. 8: 1282.

Garland-Thomson, Rosemarie. 2019. Welcoming the Unexpected. In *Human Flourishing in an Age of Gene Editing,* edited by Erik Parens and Josephine Johnston, 15–28. New York: Oxford University Press.

Ginn, Samantha L., Anais Amaya, Ian E. Alexander, Michael Edelstein, and Mohammad R. Abedi. 2018. Gene Therapy Clinical Trials Worldwide to 2017: An Update. *Journal of Gene Medicine* 20: 1–16.

Glass, Bentley. 1965. *Science and Ethical Values.* Chapel Hill: University of North Carolina Press.

Glass, Bentley. 1971. Science: Endless Horizons or Golden Age? *Science* 171: 23–29.

Greely, Henry T. 2016. *The End of Sex and the Future of Human Reproduction.* Cambridge, MA: Harvard University Press.

Gross, Neil. 2013. *Why Are Professors Liberal and Why Do Conservatives Care?* Cambridge, MA: Harvard University Press.

Haidt, Jonathan. 2007. The New Synthesis in Moral Psychology. *Science* 316: 998–1002.

Hamilton, Michael. 1972. *The New Genetics and the Future of Man.* Grand Rapids, MI: William B. Eerdmans.

Harris, John. 2007. *Enhancing Evolution: The Ethical Case for Making Better People.* Princeton, NJ: Princeton University Press.

Holtug, Nils. 1993. Human Gene Therapy: Down the Slippery Slope? *Bioethics* 7, no. 5: 402–19.

Horwitz, Allan V. 2016. *What's Normal? Reconciling Biology and Culture.* New York: Oxford University Press.

Human Gene Therapy Subcommittee, Recombinant DNA Advisory Committee. 1985. Points to Consider in the Design and Submission of Human Somatic Cell Gene Therapy Protocols. *Recombinant DNA Technical Bulletin* 8, no. 4: 181–86.

Hurlbut, J. Benjamin. 2015. Remembering the Future: Science, Law and the Legacy of Asilomar. In *Dreamscapes of Modernity: Sociotechnical Imaginaries and the Fabrication of Power*, edited by Sheila Jasanoff and Sang-Hyun Kim, 126–51. University of Chicago Press.

Huxley, Julian. 1963. Eugenics in Evolutionary Perspective. *Perspectives in Biology and Medicine* Winter 6, no. 2: 155–87.

Hynes, Richard O., Barry S. Coller, and Matthew Porteus. 2017. Toward Responsible Human Genome Editing. *JAMA* 317, no. 18: 1829–30.

Jasanoff, Sheila. 2005. *Designs on Nature: Science and Democracy in Europe and the United States.* Princeton, NJ: Princeton University Press.

Jinek, Martin, Krzysztof Chylinski, Ines Fonfara, Michael Hauer, Jennifer A. Doudna, and Emmanuelle Charpentier. 2012. A Programmable Dual-RNA-Guided DNA Endonuclease in Adaptive Bacterial Immunity. *Science*, 1225829.

Jones, David Albert. 2011. Is There a Logical Slippery Slope from Voluntary to Nonvoluntary Euthanasia? *Kennedy Institute of Ethics Journal* 21, no. 4: 379–404.

Jonsen, Albert R. 1991. Casuistry as Methodology in Clinical Ethics. *Theoretical Medicine* 12: 295–307.

Jonsen, Albert R. 1998. *The Birth of Bioethics.* New York: Oxford University Press.

Juengst, Eric T. 1997. Can Enhancement Be Distinguished from Prevention in Genetic Medicine. *Journal of Medicine and Philosophy* 22: 125–42.

Juengst, Eric T. 2017. Crowdsourcing the Moral Limits of Human Gene Editing? *Hastings Center Report* 47, no. 3: 15–23.

Juengst, Eric T., and Daniel Moseley. 2015. Human Enhancement. *The Stanford Encyclopedia of Philosophy*, http://plato.stanford.edu/archives/sum2015/entries/enhancement.

Kass, Leon R. 1972. New beginnings in life. In *The New Genetics and the Future of Man*, edited by Michael Hamilton, 15–63. Grand Rapids, MI: Eerdmans.

Keenan, James F., S.J. 2017. Openness, with caution and suspicion, about human enhancement. In *Contemporary Controversies in Catholic Bioethics*, edited by J. T. Eberl, 297–312. New York: Springer.

Kevles, Daniel. 1985. *In the Name of Eugenics: Genetics and the Uses of Human Heredity*. Berkeley: University of California Press.

Kimmelman, Jonathan. 2010. *Gene Transfer and the Ethics of First-in-Human Research*. New York: Cambridge University Press.

Lander, Eric S. 2015. Brave New Genome. *New England Journal of Medicine* 373, no. 1: 5–8.

Lanphier, Edward, Fyodor Urnov, Sarah Ehlen Haecker, Michael Werner, and Joanna Smolenski. 2015. Don't Edit the Human Germ Line. *Nature* 519, no. 7544: 410–11.

Launis, Veikko. 2002. Human Gene Therapy and the Slippery Slope Argument. *Medicine, Health Care and Philosophy* 5: 169–79.

Lerner, Barron H., and Arthur L. Caplan. 2015. Euthanasia in Belgium and the Netherlands on a Slippery Slope? *JAMA Internal Medicine* 175, no. 10: 1640–41.

Lewis, Penney. 2007. The Empirical Slippery Slope from Voluntary to Non-Voluntary Euthanasia. *Journal of Law, Medicine and Ethics* 35, no. 1: 197–210.

Lode, Eric. 1999. Slippery Slope Arguments and Legal Reasoning. *California Law Review* 87, no. 6: 1469–543.

Lowenstein, P. R. 1997. Editorial. *Gene Therapy* 4: 755–56.

Luria, S. E. 1965. Directed Genetic Change: Perspectives from Molecular Genetics. In *The Control of Human Heredity and Evolution*, edited by T. M. Sonneborn, 1–19. New York: Macmillan.

Ma, Hong, et al. 2017. Correction of a Pathogenic Gene Mutation in Human Embryos. *Nature* 548: 413–19.

Marchione, Marilynn. 2018. Chinese Researcher Claims First Gene-Edited Babies. Associated Press, November 26.

McClosky, Herbert, and John Zaller. 1984. *The American Ethos: Public Attitudes Toward Capitalism and Democracy*. Cambridge, MA: Harvard University Press.

McIntyre, Alison. 2018. Doctrine of Double Effect. *Stanford Encyclopedia of Philosophy*. https://plato.stanford.edu/entries/double-effect/.

Mehlman, Maxwell J. 2012. *Transhumanist Dreams and Dystopian Nightmares: The Promise and Peril of Genetic Engineering*. Baltimore, MD: Johns Hopkins University Press.

Meilaender, Gilbert. 1997. Begetting and Cloning. *First Things*, June/July, 41–43.

Miller, A. Dusty. 1992. Human Gene Therapy Comes of Age. *Nature* 357: 455–60.

Moore, Kelly. 1996. Organizing Integrity: American Science and the Creation of Public Interest Organizations, 1955–1975. *American Journal of Sociology* 101, no. 6, May: 1592–627.

Muller, Hermann. 1963. Genetic Progress by Voluntarily Conducted Germinal Choice. In *Man and His Future*, edited by Gordon Wolstenholme, 247–362. London: J. & A. Churchill.

Nair, Prashant. 2017. QnAs with Alta Charo and George Church. *PNAS* 114, no. 23: 5769–71.

Naldini, Luigi. 2015. Gene Therapy Returns to Centre Stage. *Nature* 526: 351–60.

National Academies of Sciences, Engineering, and Medicine. 2016. *Mitochondrial Replacement Techniques: Ethical, Social, and Policy Considerations*. Washington, DC: National Academies Press.

National Academies of Sciences, Engineering, and Medicine. 2017. *Human Genome Editing: Science, Ethics, and Governance*. Washington, DC: National Academies Press.

Nationale Akademie der Wissenschaften Leopoldina. 2015. The Opportunities and Limits of Genome Editing.

NIH Director. 2015. Statement on NIH Funding of Research Using Gene-Editing Technologies in Human Embryos.

Noble, Ray, Gulam Bahadur, Mohammad Iqbal, and Arnab Sanyal. 2008. Pandora's Box: Ethics of PGD for Inherited Risk of Late-Onset Disorders. *Reproductive BioMedicine Online* 17, no. 3: 55–60.

Parens, Erik. 1998. Is Better Always Good? The Enhancement Project. In *Enhancing Human Traits: Ethical and Social Implications*, edited by Erik Parens, 1–28. Washington, DC: Georgetown University Press.

Parens, Erik, and Josephine Johnston. 2019. *Human Flourishing in an Age of Gene Editing*. New York: Oxford University Press.

Paul, Diane B. 1995. *Controlling Human Heredity: 1865 to the Present*. Amherst, NY: Humanity Books.

Peterson, Anna L. 2001. *Being Human: Ethics, Environment, and Our Place in the World*. Berkeley: University of California Press.

Porter, Theodore M. 1995. *Trust in Numbers: The Pursuit of Objectivity in Science and Public Life*. Princeton, NJ: Princeton University Press.

Prag, Patrick, and Melinda C. Mills. 2017. Assisted Reproductive Technology in Europe: Usage and Regulation in the Context of Cross-Border Reproductive Care. In *Childlessness in Europe: Contexts, Causes, and Consequences*, edited by Michaela Kreyenfeld and Dirk Konietzka, 289–309. Cham, Switzerland: Springer.

President's Council on Bioethics. 2003. *Beyond Therapy: Biotechnology and the Pursuit of Happiness*. Washington, DC: President's Council on Bioethics.

Ramsey, Paul. 1970. *Fabricated Man: The Ethics of Genetic Control*. New Haven, CT: Yale University Press.

Ramsey, Paul. 1972. Genetic Therapy: A Theologian's Response. In *The New Genetics and the Future of Man*, edited by Michael Hamilton. Grand Rapids, MI: William B. Eerdmans.

Regalado, Antonio. 2018. Rogue Chinese CRISPR Scientist Cited US Report as His Green Light. *Technology Review*, November 27.

Richards, Sue, Nazneen Aziz, Sherri Bale, David Bick, Soma Das, Julie Gastier-Foster, Wayne W. Grody, Madhuri Hegde, Elaine Lyon, Elaine Spector, Karl Voelkerding, and Heidi L. Rehm. 2015. Standards and Guidelines for the Interpretation of Sequence Variants: A Joint Consensus Recommendation of the American College of Medical Genetics and Genomics and the Association for Molecular Pathology. *Genetics in Medicine* 17, no. 5: 405–24.

Richter, Trevor, Sandra Nestler-Parr, Robert Babela, Zeba M. Khan, Theresa Tesoro, Elizabeth Molsen, and Dyfrig A. Hughes. 2015. Rare Disease Terminology and Definitions—A Systematic Global Review: Report of the ISPOR Rare Disease Special Interest Group. *Value in Health* 18: 906–14.

Rivron, Nicolas, Martin Pera, Janet Rossant, Alfonso Martinez Arias, Magdalena Zernicka-Goetz, Jianping Fu, Susanne wan den Brink, Annelien Bredenoord, Wybo Dondorp, Guido de Wert, Insoo Hyun, Megan Munsie, and Rosario Isasi. 2018. Debate Ethics of Embryo Models from Stem Cells. *Nature* 564: 183–85.

Rizzo, Mario J., and Douglas Glen Whitman. 2003. The Camel's Nose Is in the Tent: Rules, Theories, and Slippery Slopes. *UCLA Law Review* 51: 539–92.

Rizzo, Mario J., and Douglas Glen Whitman. 2009. Little Brother Is Watching You: New Paternalism on the Slippery Slopes. *Arizona Law Review* 51: 685–739.

Robertson, John. 1985. Genetic Alteration of Embryos: The Ethical Issues. In *Genetics and the Law III*, edited by Aubrey Milunsky and George Annas, 115–27. New York: Plenum Press.

Robertson, John A. 2003. Procreative Liberty in the Era of Genomics. *American Journal of Law and Medicine* 29: 439–87.

Rosen, Christine. 2004. *Preaching Eugenics: Religious Leaders and the American Eugenics Movement*. New York: Oxford University Press.

Rothman, David J. 1991. *Strangers by the Bedside: A History of How Law and Bioethics Transformed Medical Decision Making*. New York: Basic Books.

Sandel, Michael J. 2004. The Case Against Perfection. *The Atlantic Monthly*, April.

Savulescu, Julian. 2001. Procreative Beneficence: Why We Should Select the Best Children. *Bioethics* 15: 413–26.

Schaid, Daniel J., Wenan Chen, and Nicholas B. Larson. 2018. From Genome-Wide Associations to Candidate Causal Variants by Statistical Fine-Mapping. *Nature Reviews Genetics* 19: 491–504.

Scherz, Paul. 2017. The Mechanism and Applications of CRISPR-Cas9. *The National Catholics Bioethics Quarterly,* Spring, 29–36.

Scully, Jackie Leach. 2019. Choice, Chance, and Acceptance. In *Human Flourishing in an Age of Gene Editing,* ed. Erik Parens and Josephine Johnston, 143–56. New York: Oxford University Press.

Sinsheimer, Robert L. 1969. The Prospect for Designed Genetic Change. *American Scientist* 57, no. 1: 134–42.

Solnick, Sara J., and David Hemenway. 1998. Is More Always Better?: A Survey on Positional Concerns. *Journal of Economics Behavior and Organization* 37: 373–83.

Sonneborn, Tracy M. 1965. *The Control of Human Heredity and Evolution.* New York: Macmillan.

Steffann, Julie, Pierre Jouannet, Jean-Paul Bonnefont, Herve Chneiweiss, and Nelly Frydman. 2018. Could Failure in Preimplantation Genetic Diagnosis Justify Editing the Human Embryo Genome? *Cell Stem Cell,* no. 22: 481–82.

Suter, Sonia M. 2018. The Tyranny of Choice: Reproductive Selection in the Future. *Journal of Law and Biosciences* 5, no. 2: 262–300.

Sykora, Peter, and Arthur Caplan. 2017. The Council of Europe Should Not Reaffirm the Ban on Germline Genome Editing in Humans. *EMBO Reports* 18, no. 11: 1871–72.

Thomas, Clare E., Anja Ehrhardt, and Mark A. Kay. 2003. Progress and Problems with the Use of Viral Vectors for Gene Therapy. *Nature Reviews Genetics* 4: 346–58.

UNESCO. 2015. UNESCO Panel of Experts Calls for Ban on "Editing" of Human DNA to Avoid Unethical Tampering with Hereditary Traits.

van der Burg, Wibren. 1991. The Slippery Slope Argument. *Ethics* 102, no. 1: 42–65.

Visscher, Peter M., Naomi R. Wray, Qian Zhang, Pamela Sklar, Mark I. McCarthy, Matthew A. Brown, and Jian Yang. 2017. 10 Years of GWAS Discovery: Biology, Function, and Translation. *American Journal of Human Genetics* 101: 5–22.

Volokh, Eugene. 2003. The Mechanisms of the Slippery Slope. *Harvard Law Review* 116, no. 4: 1026–137.

Wakefield, Jerome C. 1992. The Concept of Mental Disorder: On the Boundary Between Biological Facts and Social Values. *American Psychologist* 47, no. 3: 373–88.

Walters, LeRoy. 1986. The Ethics of Human Gene Therapy. *Nature* 320: 225–27.

Walters, LeRoy. 1991. Human Gene Therapy: Ethics and Public Policy. *Human Gene Therapy* 2: 115–22.

Walters, LeRoy, and Julie Gage Palmer. 1997. *The Ethics of Human Gene Therapy*. New York: Oxford University Press.

Walton, Douglas. 1992. *The Slippery Slope Argument*. New York: Oxford University Press.

Walton, Douglas. 2015. The Basic Slippery Slope Argument. *Informal Logic* 35, no. 3: 273–311.

Walton, Douglas. 2016. The Slippery Slope Argument in the Ethical Debate on Genetic Engineering of Humans. *Science and Engineering Ethics* 23, no. 6: 1507–28.

Weatherall, D. J. 1988. The Slow Road to Gene Therapy. *Nature* 331: 13–14.

Wilde, Melissa J., and Sabrina Danielsen. 2014. Fewer and Better Children: Race, Class, Religion, and Birth Control Reform in America. *American Journal of Sociology* 119, no. 6: 1710–60.

Wilper, Andrew P., Steffie Woolhandler, Karen E. Lasser, Danny McCormick, David H. Bor, and David U. Himmelstein. 2009. Health Insurance and Morality in US Adults. *American Journal of Public Health* 99, no. 12: 2289–95.

Wood, Andrew et al. 2014. Defining the Role of Common Variation in the Genomic and Biological Architecture of Adult Human Height. *Nature Genetics* 46, No. 11: 1173–86.

Woollard, Fiona. 2016. Doing vs. Allowing Harm. *Stanford Encyclopedia of Philosophy*. https://plato.stanford.edu/entries/doing-allowing/.

Wright, Susan. 1994. *Molecular Politics: Developing American and British Regulatory Policy for Genetic Engineering, 1972-1982*. Chicago: University of Chicago Press.

Zimmerman, Burke K. 1991. Human Germ-Line Therapy: The Case for Its Development and Use. *Journal of Medicine and Philosophy* 16: 593–612.

Index

For the benefit of digital users, indexed terms that span two pages (e.g., 52–53) may, on occasion, appear on only one of those pages.

Figures are indicated by *f* following the page number

Abbott, Andrew, 154n14
abortion, 11–12, 50–51, 52, 84, 107, 145–46, 147, 169–70n42
adenosine deaminase (ADA) deficiency, 37–38, 43, 53
Agar, Nicholas, 124
Alliance for Regenerative Medicine, 71–72
American Society for Gene and Cell Therapy, 71
Anderson, W. French, 34, 36–37, 47, 58–59, 160n31
autonomy, value of, 22, 46–47, 65, 87, 98, 107–8, 112, 126, 127, 128–29, 135, 136–37, 140

Baltimore, David, 74
barrier anchoring authority debate, 142–43
barrier creation debate, 140–42
barriers, combinations of, 33*f*, 33–34
barriers downslope of median trait barrier (hypothetical)
    family genes barrier, 115–22
    family genes barrier, design of, 118–20
    family genes barrier, values imperiling, 122
    family genes barrier, values supporting, 120–22
    human barrier, boundary of, 122
    human barrier, design of boundary of, 124–25
    human barrier, values imperiling boundary of, 126
    human barrier, values supporting boundary of, 125
    liberal eugenics barrier, 126–28
    liberal eugenics barrier, values supporting, 130–31
    liberal eugenics barrier, weaknesses of, 128–30
    medicine barrier, goals of, 113–14
    medicine barrier, weaknesses of goals of, 114–15
    overview, 23–24, 111–13, 131–32
barrier strengths
    design strength of disease/enhancement barrier, 35–38
    design strength of somatic/germline barrier, 39–40
    hypothetical strengths of somatic/germline barrier, 20–23, 21*f*
    two barriers in use, 42–43
    values supporting disease/enhancement barrier, 38–39
    values supporting the somatic/germline barrier, 40–42
beneficence, value of, 22, 31, 32, 34, 38–39, 45, 46–50, 52, 58, 65, 77–78, 80, 87, 98, 101, 106–7, 109–10, 111, 112, 122, 126, 127, 134, 135, 136–37, 140–41, 146, 168n35
Berg, Paul, 74

beta thalassemia, 53, 68, 168–69n38
Bostrom, Nick, 144–45
A Brave New World (Huxley), 14, 73, 85, 133, 143–44
Brock, Dan W., 18–19, 101–2
Buchanan, Allen, 18–19, 101–2
Burg, B. van der, 154–55n19

cancer, immune cell modification and, 63–64
Caplan, Arthur, 6, 169–70n42
casuistry, 17, 156n33
category boundary re-appraisal mechanism, 156n31
Center for Genetics and Society, 72, 102–3, 165–66n16
change in participants and their values, 44–46
Charo, R. Alta, 74
Charpentier, Emmanuelle, 69
Chinese Academy of Science, 75
Chinese HIV-resistant twins, 1, 77, 86, 111, 138–39
Christian theologians, 29
Church, George, 169–70n42, 170n44
comparative analysis speculations, 147–49
Conley, Dalton, 96–97
continuity vagueness
  Anderson on, 59
  barrier vulnerability and, 37
  family genes barrier and, 118–19
  functional compromise and, 155n29, 172n60
  future barriers and, 138
  HGE debate and, 15–16, 145, 147
  liberal eugenics barrier and, 128–29
  median trait barrier and, 99
  normalcy and, 63
  prevalent variant barrier and, 91–92
  safety and, 74, 75–76
  strong barriers and, 20
cosmetic bodily enhancements, 139
Crick, Francis, 3, 28
CRISPR-CAS9 system
  development of, 4–5, 69
  future polygenic selection/ modification and, 94–97
  mosaicism and, 165n13
  technology of, 69
  use of, 70, 108–9
  See also Charpentier, Emmanuelle CRISPR era; Doudna, Jennifer; median trait barrier
CRISPR Era
  barriers at beginning of, 69–77
  barriers examined, 108–9
  ethics of HGE in, 76
  overview, 23–24, 67–69
  See also CRISPR-CAS9 system; median trait barrier; NASEM barrier (proposed); NASEM report; prevalent variant barrier
cystic fibrosis, 35–36, 60, 168–69n38

Daniels, Norman, 18–19, 61, 101–2
Davis, Bernard, 30–32, 36–37
dehumanization, 14–15, 112
design of prevalent variant barrier, amendment to, 92–94
design strength of barrier, amendment to design of prevalent variant barrier, 92–94
disease (term), instability of, 85–87
disease/enhancement barrier
  change in participants and their values, 44–46
  design strength of, 35–38
  disease-causing mutations, 168n37
  early HGE debate on, 29–35, 33*f*
  at end of 20th century, 65–66

INDEX 193

NASEM barrier (proposed)
and, 85–87
in use, 42–43
values supporting, 38–39
disease/enhancement barrier
weakening
disease as deviation from
normalcy, 60–62
overview, 57–60, 135–36
susceptibility of barrier anchored
at normalcy, 62–65
DNA (deoxyribonucleic acid)
discovery of, 3, 28
gene therapy and, 32–33
modification of, 71–72, 165n12
Dobzhansky, Theodosius, 41
double effect doctrine, 161–62n53
Doudna, Jennifer, 5, 69, 74

editing, term usage, 4–5
embryo destruction, 169–70n42
enhancement
defined, 177–78n12
early HGE debate on, 33f
NASEM barrier (proposed)
and, 85–87
prevalent variant barrier
and, 87–89
enhancements
defined, 31
scientific innovations and, 135–36
See also disease/enhancement
barrier
ethics, safety and, 165–66n16
eugenics
as bottom of the slope, 14–15
dystopian views of, 14–15
early debates on, 3–4, 28, 134, 144
early mainline movement,
2–3, 25–27
enhancement and, 28, 31–32, 35–
36, 39, 40–41, 135–36
genetic diseases and, 28, 34

genetic load, 27–28, 31
HGE debate and, 3–4, 28
high-tech consumer, 72, 102–3
International Bioethics
Committee of UNESCO on, 73
modern version of, 15, 165–66n16
Morison, R. S., 95
Nazi race-based eugenics, 3, 5, 27,
133, 134
personal level, 59
post-WWII concerns, 27–28
race-based, 27
right to negative eugenics, 51
species perfecting and, 29–30, 33–
34, 39, 48, 134, 137
species progress and, 28
transhumanists, 98, 144–45
value of human control, 42
See also liberal eugenics barrier;
reform eugenicists
euthanasia, 10, 11–12, 147, 149,
156–57n37

fairness
alternate distribution values
and, 174n72
See also justice (fairness), value of
family genes barrier
design of, 118–20
germline modification and, 139
overview, 115–18
values imperiling, 122
values supporting, 120–22
first barriers in HGE debate
barrier strengths, 35–43
change in participants and their
values, 44–46
disease/enhancement barrier (end
of 20th century), 65–66
disease/enhancement barrier
weakening, 57–65
first barriers, 29–35, 33f
overview, 25–29

first barriers in HGE debate (*Cont.*)
  somatic/germline barrier (end of 20th century), 65–66
  somatic/germline barrier weakening, 46–57
Fletcher, John, 36–37, 47, 97
forced sterilization, 2–3, 26, 27–28
Friedmann, Theodore, 32, 35–36, 42–43, 48–49
functional compromise, 155n29, 172n60
future barriers on HGE slope, 137–38
future polygenic selection/modification, NASEM report and, 94–97

Galton, Francis, 25–26
Gattaca (1997 film) (Niccol), 14–15, 79, 85, 87, 117–18, 133, 141–42, 143–44, 165–66n16
gene replacement, 163n69
gene therapy
  DNA (deoxyribonucleic acid) and, 32–33
  risk/benefit ratio, 160n31, 160n32
  single-gene (monogenic) traits and, 136, 159–60n30, 163n75
  terminology usage, 32–33
  See also somatic gene therapy
genetic diseases
  early HGE attempts, 4, 10–11, 64–65
  eugenics and, 28, 29–30
  monogenic (single-gene), 31–32, 35–36
  PGD and, 53–54
  somatic gene therapy, 13, 42–43
  traits of, 35–36
  See also adenosine deaminase (ADA) deficiency; beta thalassemia; disease/enhancement barrier; sickle cell anemia; Tay-Sachs disease
genetic engineering (term), 32–33
genetic load, 27–28, 31
genetic relatedness, 168n35
gene variants, 54–55, 90–91, 96, 97, 104, 114, 119–20, 139–40, 150, 173n62, 173n63
genome-wide association studies (GWAS), 96–97, 172n61
germline barrier
  autonomy emphasis undermining, 50–52
  beneficence emphasis undermining, 46–50
  during CRISPR era, 70
  defense of, 99, 102–3
  early HGE debate on, 33f
  first similarity vagueness mechanism, 80–82
  further downslope of, 111–13
  future polygenic selection/modification and, 94–97
  limited number of cases, 82–83
  moral defense of, 18–20
  NASEM on, 165n12
  PGD and, 52–55, 137
  prevalent variant barrier and, 90
  safety and, 148, 171n53
  second similarity vagueness mechanism, 83–85
  slippery slope across, 78–80, 137
  as species modifying, 20
  US scientific establishment questions need for, 73–77
  See also NASEM report; somatic/germline barrier
germline modification
  Center for Genetics and Society on, 165–66n16
  development of, 6
  fairness and, 87–88
  family genes barrier and, 118–19, 139
  general public and, 150–52
  germline selection and, 94–97

INDEX 195

NASEM barrier (proposed)
  and, 85–87
  NASEM on, 165n12
  PGD and, 54–55, 108–9
  prevalent variant barrier and,
    88–89, 90
  renewal of interest in, 70–71
  reproductive genetics
    and, 105–6
  support of, 102–3
  as target, 16
  waning of interest in, 67–69
germline selection, 54–55, 109
germline modification and,
  94–97, 162n62
Glass, H. Bentley, 29–30
God's will
  decline in values following, 46,
    112, 136–37
  disease/enhancement barrier
    and, 39
  family genes barrier and,
    121–22, 140
  germline barrier and,
    19–20, 73–74
  humanity barrier and, 125
  normalcy and, 62–63
  PGD and, 56, 137
  somatic/germline barrier and,
    40–42, 135
Greely, Henry, 74
GWAS (genome-wide association
  studies), 96–97, 172n61

Haidt, Jonathan, 14–15, 112
harm, types of, 129–30
Harris, John, 18–19, 127–28
heritable genome editing, 79, 81–82,
  168n34, 170n47
HGE (human gene editing) debate
  barriers, 18–24, 21f
  overview, 1–5
  public bioethical debate, 5–9
  slippery slopes, 9–11

slippery slopes as micro-structure
  of HGE debate, 15–18
slope as micro-structure
  of, 11–15
historical barriers on HGE
  slope, 134–37
HIV-resistant twins, 1, 77, 86,
  111, 138–39
Holmes, Oliver Wendell, 26
Holtug, Nils, 154–55n19
human barrier
  boundary of, 122
  design of boundary of, 124–25
  values imperiling boundary
    of, 126
  values supporting boundary
    of, 125
Human Fertilisation and
  Embryology Authority, 143
human gene editing
  barrier anchoring authority
    debate, 142–43
  barrier creation debate, 140–42
  comparative analysis
    speculations, 147–49
  future barriers on HGE
    slope, 137–38
  future possible actions, 150–52
  historical barriers on HGE
    slope, 134–37
  objectivity and strong
    barriers, 149–50
  overview, 133
  slippery slopes (HGE case),
    general conclusions,
    23–24, 145–47
  slope direction rejection, 143–45
  See also HGE (human gene
    editing) debate
Human Gene Therapy Subcommittee
  of the RAC, 160n32
human genetic engineering (HGE),
  defined, 1
human genetic inheritance, 1–3

humanity barrier, future possible actions and, 139–40
human reproduction
 forced sterilization, 2–3
 modification to, 2–3
Huntington's disease, 60, 169n41
Huxley, Aldous, 14
Huxley, Julian, 28, 40–41

immune cell modification, 63–64
International Bioethics Committee of UNESCO, 73
in vitro fertilization (IVF), 50–51, 54, 55, 79–80, 84, 116

Japan Society of Gene Therapy, 71
Juengst, Eric, 58, 113, 114–15, 141, 142
justice (fairness), value of, 45, 46–47, 87–88, 89, 91–92, 93, 98, 100, 101–2, 103–6, 109–10, 112, 120–21, 135, 136–37, 138

Kass, Leon, 14
Kimmelman, Jonathan, 160n33

Lander, Eric, 104, 122–23, 169n41
liberal eugenics barrier
 as final barrier, 140
 overview, 126–28, 132
 values supporting, 130–31
 weaknesses of, 128–30
 See also liberal eugenics barrier
Luria, Salvador, 44
lysosomal storage diseases, 168n37, 168–69n38

macro-/micro-structure analogy, 8–9
"made by"/"designed by" idea, 178n20
median trait barrier, 87–89

alternate distribution values, 174n72
 challenges to designing, 99–100
 designing of, 97–100
 overview, 97–100
 values imperiling, 103–7
 values supporting, 100–3
medicalization, 114–15
medicine barrier
 goals, 113–14
 weaknesses of goals, 114–15
Meilaender, Gilbert, 56
Mendelian inheritance, 1–2, 35–36, 88–89, 159n28, 171n55
Mill, John Stuart, 128
monogenic (single-gene) traits. *See* single-gene (monogenic) traits
Moore, Kelly, 29
morality
 abortion, 169–70n42
 common morality, 157n38
 embryo destruction, 169–70n42
 moral firebreaks, 18–19
Morison, R. S., 95
mosaicism, 71, 165n13
Moseley, Daniel, 114–15
mutations, disease-causing, 168n37
mutations, recessive, 168n37, 168–69n38

NASEM report
 advocating replacing somatic/germline barrier, 23–24, 77–85, 114–15, 137
 design strength of barrier, 90–94
 on enhancement, 103, 170n48, 171n57
 family genes barrier, 122
 future polygenic selection/modification, 94–97
 on germline barrier, 108–9, 119, 122
 knowledge increase and, 148

NASEM barrier (proposed),
  85–87, 90–94
  overview, 76–77
  prevalent variant barrier and,
    87–90, 92–94
  proposed barrier, 85–87
  on secondary harms, 130
  somatic/germline barrier and,
    77–85, 137
  US government regulations based
    on, 112–13
  on value of justice/fairness,
    103, 109
National Academies of Science,
  Engineering, and Medicine
  (NASEM), 7, 61, 71, 75,
  165n12, 167n25
  See also NASEM barrier
    (proposed); NASEM report
National Institutes of Health
  (NIH), 44–45, 70–71, 76–77,
  99–100, 152
nature/God's will. *See* God's will
Nazi race-based eugenics, 3, 5, 27,
  133, 134
non-maleficence (avoiding harm),
  20–22, 39, 40, 45, 46–47, 70, 71,
  72–73, 74, 75–76, 77–78, 87,
  101, 108–9, 112, 127, 128–29,
  130–31, 135, 136–37
non-maleficence (safety), 126

objectivity and strong
  barriers, 149–50
O'Donovan, Oliver, 56
1000 Genomes Project, 173n62

Parens, Erik, 62–63, 113
PGD (pre-implantation
  genetic diagnosis). *See*
  pre-implantation genetic
  diagnosis (PGD)
polygenic traits

disease/enhancement barrier and,
  36–37, 39, 60, 64–65
engineering debate and, 31–32
family genes barrier and, 118
future polygenic selection/
  modification, 94–97
identification of, 6
median trait barrier and, 97–100,
  109–10, 112–13, 138
risk scores, 100, 104, 115–16,
  174–75n73
pre-implantation genetic
  diagnosis (PGD)
  advocating replacing somatic/
    germline barrier and, 94
  during CRISPR era, 68–69
  debates on, 53–54
  first similarity vagueness
    mechanism and, 80–82
  hypothetical new barrier
    downslope, 54–57
  invention of, 53
  limited number of cases, 82–83
  recessive diseases and, 169n41
  second similarity vagueness
    mechanism and, 83–85
  selection through, 68–69, 79–80,
    81, 97, 174n71
  WGS-PGD users, 178n13
prevalent variant barrier, 174–75n73
  amendment to design of, 92–94
  design strength of barrier, 90–94
  gene variants, 173n62, 173n63
  locating on the slope, 89–90
  meaning of, 172n61
  overview, 87–90
public bioethical debate
  HGE debate, 5–9
  participants in, 18
  resistance to radical change, 22–23
  slippery slopes and, 9–11
  slope as micro-structure of, 11–15

Ramsey, Paul, 29, 34–35, 40, 41–42
Rawls, John, 101–2
recessive diseases, 168–69n38, 169n41
recessive mutations, 168n37
Recombinant DNA Advisory Committee (RAC), 159–60n30
reform eugenicists, 28, 34, 39, 40–42, 98
risk/benefit ratio, 6, 37–38, 43, 68–69, 87–88, 103, 112, 160n31
risks, data on, 167n30
Robertson, John, 51, 108, 128
Roblin, Richard, 32, 35–36
Royal Society (UK), 75

Sandel, Michael J., 56
Savulescu, Julian, 107–8
selection, modification and, 162n62
sickle cell anemia
  environment and, 93
  gene editing and, 2
  homozygous parents, 119–20, 168–69n38
  NASEM barrier (proposed) and, 83–84, 85, 119, 137
  PGD and, 53–55, 68–69, 79–80, 81, 137
  single-gene (monogenic) trait of, 95, 134, 137
similarity vagueness
  single-gene (monogenic) traits and, 64–65
  terrain and, 16, 112–13, 119, 147
similarity vagueness mechanism, first, 59, 80–82
similarity vagueness mechanism, second, 59, 83–85
single-gene (monogenic) traits
  gene therapy and, 136, 159–60n30, 163n75
  PGD and, 53, 55, 68–69, 94, 97
  prevalent variant barrier and, 95, 137, 138
  recessive diseases, 4–5, 134
  similarity vagueness and, 64–65
Sinsheimer, Robert, 3–4
slippery slope arguments, 154–55n19
  attitude-altering mechanisms, 11
  criticism of, 156–57n37
  empirical/legitimate, 10
  legislative mechanisms, 10–11
  logical/conceptual, 10
  psychological processes and, 156n31
  technical knowledge mechanisms, 11
"The Slippery Slope Argument" (Burg), 154–55n19
slippery slopes
  as micro-structure of HGE debate, 15–18
  overview, 9–11
slippery slopes (HGE case), general conclusions, 145–47
slope
  as micro-structure of HGE debate, 11–15
  prevalent variant barrier on, 89–90
slope direction rejection, 143–45
Social Darwinism, 8, 27–28
social engineering, 2
somatic gene therapy, 13, 15, 35, 36–38, 63–64, 71, 74
  See also somatic/germline barrier
somatic/germline barrier
  biological basis for, 149–50
  change in participants and their values, 44–46
  constructing barriers below, 120
  design strength of, 39–40
  early HGE debate on, 29–35, 33f, 134

emphasis on autonomy
undermines germline
barrier, 58–59
at end of 20th century, 65–66
first similarity vagueness
mechanism, 59, 80–82
hypothetical strengths of,
20–23, 21*f*
limited number of cases, 82–83
NASEM report advocating
replacing of, 77–85, 122, 137
prevalent variant barrier
and, 90–91
second similarity vagueness
mechanism, 83–85
slippery slope across, 78–80
in use, 42–43
value of autonomy and, 146–47
values supporting, 40–42, 136
somatic/germline barrier
weakening
emphasis on autonomy
undermines germline
barrier, 50–52
emphasis on beneficence
undermines germline
barrier, 46–50
hypothetical new barrier
downslope, 54–57
overview, 46, 135
technology changing terrain
and, 52–54
species progress
autonomy emphasis and, 52–54
beneficence emphasis and, 46–50
early HGE debate and, 13–14
eugenicists and, 2–3, 28, 29–30,
33–34, 144–45
germline and, 20, 31, 34–35
historical and future barriers
and, 134–40
median trait barrier and, 98
normalcy and, 170n48

PGD and, 55
reform eugenicists and, 39–40
use of animal species for,
122, 139–40
values and, 40–42
Steffann, Julie, 169–70n42
Sykora, Peter, 169–70n42

target, defined, 13
Tay-Sachs disease, 13, 50–51, 57–58,
61, 63, 64–65, 85–86, 168n37
terminology usage, 32–34, 33*f*
terrain
continuity vagueness and, 15–16,
37, 119, 147
defining, 11–12, 13, 145–46
humanity barrier and, 140
liberal eugenics barrier and,
128, 140
normalcy and, 61–62
prevalent variant barrier and,
89, 90–91
similarity vagueness and, 16, 112–
13, 119, 147
strong barriers and, 20, 134–35
technology changing,
52–54, 68–69
as trait and target, 13, 34, 35–36,
39–40, 50, 145–46
values and, 22, 34, 38, 46, 50, 103,
146, 147
theologians
on embryonic life, 53, 84,
169–70n42
germline barrier and, 19–20, 42,
49–50, 170n44
as participants, 44
Protestant, 8
values and, 18, 45–46
See also God's will; Ramsey, Paul
trait (term), 171n54
transhumanists, 98, 144–45
transparency, 73, 76, 167n29

Universal Declaration on Bioethics and Human Rights of 2005, 73

values
  bioethics and public's, 157n38
  doing vs. allowing distinction, 162n62
  of eugenicists, 42
  sources of, 18
  species progress, 39, 40–42
  supporting median trait barrier, 100–3
  of theologians, 42
  value of autonomy, 107–8
  value of beneficence, 106–7
  value of human control, 42
  value of justice, 103–6
  See also autonomy, value of

variants, gene, 173n63
  See also gene variants

Wakefield, Jerome C., 164n80
Walters, LeRoy, 42–43, 48, 54, 60, 163n69
Watson, James, 3, 28
WGS (whole genome sequencing), 115–16, 118–19
whole genome sequencing (WGS), 115–16, 118–19
Wikler, Daniel, 18–19, 101–2
Williams, Granville, 18–19
World Health Organization (WHO), 114

Zimmerman, Burke K., 54